# SCIENCE,
# EVOLUTION,
## AND
# CREATIONISM

NATIONAL ACADEMY OF SCIENCES
INSTITUTE OF MEDICINE

THE NATIONAL ACADEMIES PRESS
Washington, D.C.

THE NATIONAL ACADEMIES PRESS • 500 Fifth Street, N.W. • Washington, D.C. 20001

NOTICE: The project that is the subject of this report was approved by the Council of the National Academy of Sciences, whose members are drawn from the National Academy of Sciences. The members of the committee were chosen for their special competences and with regard for appropriate balance.

Funding for this project was provided by the Council of the National Academy of Sciences, with additional support from the Christian A. Johnson Endeavor Foundation and the Biotechnology Institute. The opinions, findings, conclusions, and recommendations expressed in this report are those of the authoring committee and of the National Academy of Sciences and do not necessarily reflect the views of the external organizations that provided support.

Library of Congress Cataloging-in-Publication Data

Science, evolution, and creationism / National Academy of Sciences and Institute of Medicine of the National Academies.
    p. cm.
 Includes bibliographical references.
 ISBN-13: 978-0-309-10586-6 (pbk.)
 ISBN-10: 0-309-10586-2 (pbk.)
 ISBN-13: 978-0-309-10587-3 (pdf)
 ISBN-10: 0-309-10587-0 (pdf)
 1. Evolution (Biology) 2. Creationism. 3. Science. I. National Academy of Sciences (U.S.) II. Institute of Medicine (U.S.)
QH366.2.S35 2007
576.8—dc22

2007015904

Additional copies are available from:

The National Academies Press
500 Fifth Street, N.W.
Box 285
Washington, D.C. 20055
800/624-6242
202/334-3313 (in the Washington Metropolitan Area)
<http://www.nap.edu>

Suggested citation:
National Academy of Sciences and Institute of Medicine (2008). *Science, Evolution, and Creationism.* Washington, D.C.: The National Academies Press.

The links to websites that provide additional information to users of this book were operative as of January 3, 2008. Changes to websites and relocated information may render some links inoperative in the future.

**COMMITTEE ON REVISING** *SCIENCE AND CREATIONISM:*
*A VIEW FROM THE NATIONAL ACADEMY OF SCIENCES*

**Francisco J. Ayala**, *Chair,* University of California, Irvine*

**Bruce Alberts,** University of California, San Francisco*

**May R. Berenbaum,** University of Illinois, Urbana-Champaign*

**Betty Carvellas,** Essex High School (Vermont)

**Michael T. Clegg,** University of California, Irvine*‡

**G. Brent Dalrymple,** Oregon State University*

**Robert M. Hazen,** Carnegie Institution of Washington

**Toby M. Horn,** Carnegie Institution of Washington

**Nancy A. Moran**, University of Arizona*

**Gilbert S. Omenn**, University of Michigan†

**Robert T. Pennock,** Michigan State University

**Peter H. Raven,** Missouri Botanical Garden*

**Barbara A. Schaal,** Washington University of St. Louis*‡

**Neil deGrasse Tyson,** American Museum of Natural History

**Holly Wichman**, University of Idaho

* Member, National Academy of Sciences
† Member, Institute of Medicine
‡ Member, Council of the National Academy of Sciences

*Staff*

**Jay B. Labov,** Senior Advisor for Education and Communications,
National Academy of Sciences, and Center for Education, Division on Behavioral
and Social Sciences and Education, National Research Council

**Barbara Kline Pope,** Executive Director, National Academies Office of
Communication and National Academies Press

**Terry K. Holmer,** Senior Project Assistant, Center for Education

**B. Ashley Zauderer,** Christine A. Mirzayan Policy Fellow of the National
Academies

*Consultants*

**Steve Olson,** Bethesda, Maryland

**Edward Maibach,** George Mason University

The National Academy of Sciences is a private, nonprofit, self-perpetuating society to which distinguished scholars are elected for their achievements in research, and is dedicated to the furtherance of science and technology and to their use for the general welfare. Upon the authority of the charter granted to it by the Congress in 1863, the Academy has a mandate to advise the federal government on scientific and technical matters. The Institute of Medicine was established in 1970 by the National Academy of Sciences as both an honorific and a policy research organization, to which members are elected on the basis of their professional achievement and commitment to service in the examination of policy matters pertaining to the health of the public.

The National Academy of Sciences and the Institute of Medicine are each governed by an elected council. The NAS Council is responsible for honorific aspects of the NAS and for the corporate management of the organization. The IOM Council oversees the study activities of the Institute, as well as matters pertaining to the IOM membership. The members of both councils reviewed, revised, and approved this document.

# SCIENCE,
## EVOLUTION,
### AND
# CREATIONISM

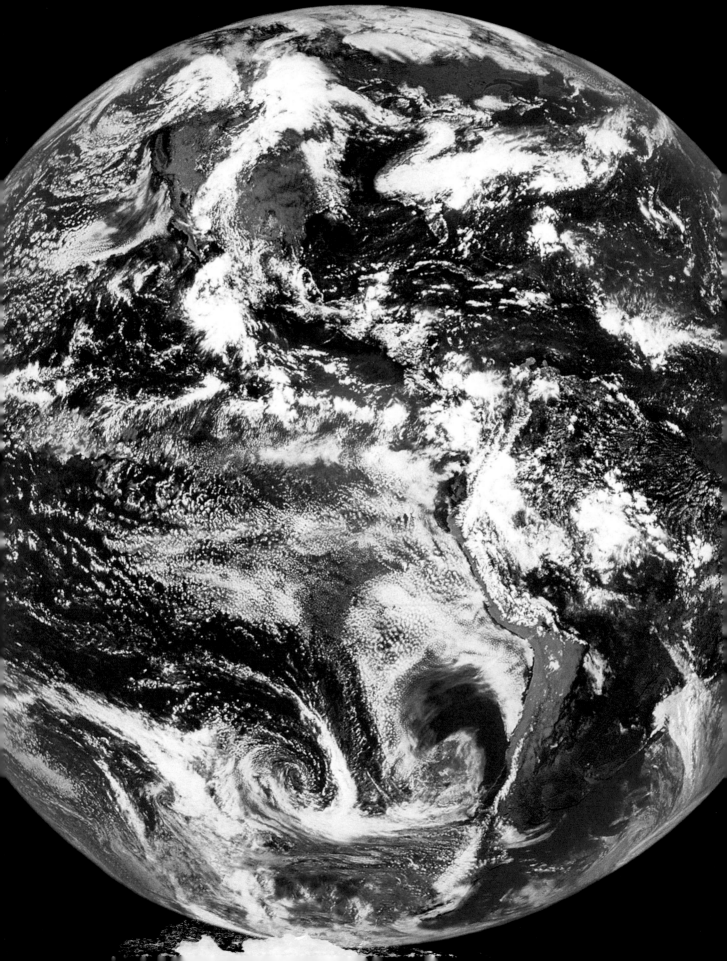

# Contents

# *Preface*

Scientific and technological advances have had profound effects on human life. In the 19th century, most families could expect to lose one or more children to disease. Today, in the United States and other developed countries, the death of a child from disease is uncommon. Every day we rely on technologies made possible through the application of scientific knowledge and processes. The computers and cell phones which we use, the cars and airplanes in which we travel, the medicines that we take, and many of the foods that we eat were developed in part through insights obtained from scientific research. Science has boosted living standards, has enabled humans to travel into Earth's orbit and to the Moon, and has given us new ways of thinking about ourselves and the universe.

Evolutionary biology has been and continues to be a cornerstone of modern science. This booklet documents some of the major contributions that an understanding of evolution has made to human well-being, including its contributions to preventing and treating human disease, developing new agricultural products, and creating industrial innovations. More broadly, evolution is a core concept in biology that is based both in the study of past life forms and in the study of the relatedness and diversity of present-day organisms. The rapid advances now being made in the life sciences and in medicine rest on principles derived from an understanding of evolution. That understanding has arisen both through the study of an ever-expanding fossil record and, equally importantly, through the application of modern biological and molecular sciences and technologies to the study of evolution. Of course, as with any active area of science, many fascinating questions remain, and this booklet highlights some of the active research that is currently under way that addresses questions about evolution.

However, polls show that many people continue to have questions about our knowledge of biological evolution. They may have been told that scientific understanding of evolution is incomplete, incorrect, or in doubt. They may be skeptical that the natural process of biological evolution could have produced such an incredible array of living things, from microscopic bacteria to whales and redwood trees, from simple sponges on coral reefs to humans capable of contemplating life's history on this planet. They may wonder if it is possible to accept evolution and still adhere to religious beliefs.

This publication speaks to those questions. It is written to serve as a resource for people who find themselves embroiled in debates about evolution. It provides information about the role that evolution plays in modern biology and the reasons why only scientifically based explanations should be included in public school science courses. Interested readers may include school board members, science teachers and other education leaders, policy makers, legal scholars, and others in the community who are committed to providing students with quality science education. This booklet is also directed to the broader audience of high-quality school and college students as well as adults who wish to become more familiar with the many strands of evidence supporting evolution and to understand why evolution is both a fact and a process that accounts for the diversity of life on Earth.

This booklet also places the study of evolution in a broader context. It defines what "theory" means in the scientific community. It shows how evolutionary theory reflects the nature of science and how it differs from religion. It explains why the overwhelming majority of the scientific community accepts evolution as the basis for modern biology. It shows that some individual scientists and religious organizations have described how, for them, evolution and their faith are not in opposition to each other. And it explains why nonscientific alternatives to evolution such as creationism (including intelligent design creationism) should not be part of the science curriculum in the nation's public schools.

*Science, Evolution, and Creationism* is the third edition of a publication first issued in 1984 by the National Academy of Sciences, an independent society of scientists elected by their peers for outstanding contributions to their field. The National Academy of Sciences has had a mandate from Congress since 1863 to advise the federal government on issues of science and technology. Given the increasing importance of evolution to the life, physical, and medical sciences and to the improvement of health care, this new edition is a joint publication of the National Academy of Sciences and the Institute of Medicine. The Institute of Medicine was chartered in 1970 as a component of the National Academy of Sciences to provide science-based advice on matters of biomedical science, medicine, and health.

Much has happened in evolutionary biology since the release of the first two editions of this booklet, and this new edition provides important updates about these developments. Fossil discoveries have continued to produce new and compelling evidence about evolutionary history. New information and understanding about the molecules that make up organisms has emerged, including the complete DNA sequences of humans. DNA sequencing has become a powerful tool for establishing genetic relationships among species. DNA evidence has both confirmed fossil evidence and allowed studies of evolution where the fossil record is still incomplete. An entirely new field, evolutionary developmental biology, enables scientists to study how the genetic changes that have occurred throughout history have shaped the forms and functions of organisms. The study of biological evolution constitutes one of the most active and far-reaching endeavors in all of modern science.

The public controversies that swirl around evolution also have changed. In the 1980s many people opposed to the teaching of evolution in public schools supported legislation that would have required biology teachers to discuss "scientific creationism" — the assertion that the fossil record and the planet's geological features are consistent with Earth and its living things being created just a few thousand years ago. Major court cases — including a Supreme Court case in

1987— ruled that "creation science" is the product of religious convictions, not scientific research, and that it cannot be taught in public schools because to do so would impose a particular religious perspective on all students.

Since then, the opponents of evolution have taken other approaches. Some have backed the view known as "intelligent design," a new form of creationism based on the contention that living things are too complex to have evolved through natural mechanisms. In 2005 a landmark court case in Dover, Pennsylvania, deemed the teaching of intelligent design unconstitutional, again because it is based on religious conviction and not science.

Others have argued that science teachers should teach the "controversies" surrounding evolution. But there is no controversy in the scientific community about whether evolution has occurred. On the contrary, the evidence supporting descent with modification, as Charles Darwin termed it, is both overwhelming and compelling. In the century and a half since Darwin, scientists have uncovered exquisite details about many of the mechanisms that underlie biological variation, inheritance, and natural selection, and they have shown how these mechanisms lead to biological change over time. Because of this immense body of evidence, scientists treat the occurrence of evolution as one of the most securely established of scientific facts. Biologists also are confident in their understanding of how evolution occurs.

This publication consists of three main chapters. The first chapter briefly describes the process of evolution, the nature of science, and differences between science and religion. The second chapter examines in greater detail the many different kinds of scientific evidence that support evolution, including evidence from fields as diverse as astronomy, paleontology, comparative anatomy, molecular biology, genetics, and anthropology. The third chapter examines several creationist perspectives, including intelligent design, and discusses the scientific and legal reasons against teaching creationist ideas in public school science classes. A selection of frequently asked questions follows the main text. "Additional Readings" include papers referenced in this booklet and other publications about evolution, the nature of science, and religion.

As *Science, Evolution, and Creationism* makes clear, the evidence for evolution can be fully compatible with religious faith. Science and religion are different ways of understanding the world. Needlessly placing them in opposition reduces the potential of each to contribute to a better future.

Ralph J. Cicerone
*President*
National Academy of Sciences

Harvey V. Fineberg
*President*
Institute of Medicine

Francisco J. Ayala
*Committee Chair*

# Acknowledgments

The preparation of *Science, Evolution, and Creationism* was supported with funds from the Council of the National Academy of Sciences and from the Christian A. Johnson Endeavor Foundation, New York, N.Y.

Representatives of this booklet's intended audiences informally reviewed this booklet prior to the final review process. Support for obtaining input from intended audiences was provided by an informal coalition of some 30 scientific societies based in the Washington, D.C., metropolitan area, the Presidents' Circle Communications Initiative of the National Academies, the Council of the National Academy of Sciences, the Biotechnology Institute (Arlington, Va.), and contributions to the National Academy of Sciences from several individual donors.

This booklet has been formally reviewed in draft form by individuals chosen for their diverse perspectives and technical expertise, in accordance with procedures approved by the Council of the National Academy of Sciences. The purpose of this independent review is to provide candid and critical comments to assist the institution in making its published report as sound as possible. The review comments and draft manuscript remain confidential to protect the integrity of the process. We thank the following individuals for their review of this report:

**Joyce Appleby,** Professor Emerita, Department of History, University of California, Los Angeles

**Constance Bertka,** Director, Dialog on Science, Ethics, and Religion, American Association for the Advancement of Science

**Donald M. Berwick,** President and CEO, Institute for Healthcare Improvement

**John I. Brauman,** J. G. Jackson–C. J. Wood Professor, Department of Chemistry, Stanford University

**Vicki L. Chandler,** Director, BIO5 Institute, University of Arizona

**Harvey V. Fineberg,** President, Institute of Medicine

**Jerry P. Gollub,** J. B. B. Professor in the Natural Sciences and Professor of Physics, Haverford College

**Susan Gottesman,** Chief, Biochemical Genetics Section, Laboratory of Molecular Biology, National Cancer Institute, National Institutes of Health

**Margaret G. Kivelson,** Distinguished Professor of Space Physics, Department of Earth and Space Sciences and Institute of Geophysics and Planetary Physics, University of California, Los Angeles

# CHAPTER ONE

# EVOLUTION AND THE NATURE OF SCIENCE

*The scientific evidence supporting biological evolution continues to grow at a rapid pace.*

For more than a century and a half, scientists have been gathering evidence that expands our understanding of both the fact and the processes of biological evolution.  They are investigating how evolution has occurred and is continuing to occur.

In 2004, for example, a team of researchers made a remarkable discovery.  On an island in far northern Canada, they found a four-foot-long fossil with features intermediate between those of a fish and a four-legged animal.  It had gills, scales, and fins, and it probably spent most of its life in the water.  But it also had lungs, a flexible neck, and a sturdy fin skeleton that could support its body in very shallow water or on land.

Earlier scientific discoveries of fossilized plants and animals had already revealed a considerable amount about the environment in which this creature lived.  About 375 million years ago, what is now Ellesmere Island in Nunavut Territory, Canada, was part of a broad plain crossed by many meandering streams.  Trees, ferns, and other ancient plants grew on the banks of the streams, creating a rich environment for bacteria, fungi, and simple animals that fed on decaying vegetation.  No large animals yet lived on the land, but Earth's oceans contained many **species** of fish, and some of those species fed on the plants and animals in shallow freshwater streams and swamps.

**[Species:** *In sexually reproducing organisms, species consist of individuals that can interbreed with each other.*]

Paleontologists searched this valley in Nunavut, near the Arctic Circle in north central Canada, for fossils when they learned that it contained sedimentary rocks deposited during the period when limbed animals were first starting to live on land. Fossils of *Tiktaalik* were discovered on the dark outcropping of rock on the right side of this photograph.

**Paleontologists** had previously found the fossils of some of these shallow-water fishes. The bones in their fins were sturdier and more complex than in other fish species, perhaps allowing them to pull themselves through plant-filled channels, and they had primitive lungs as well as gills. Paleontologists had also found, in somewhat younger sediments, fossils of fishlike animals that likely spent part of their time on land. Known as early tetrapods (a word referring to their four legs), they had modified front and back fins that resembled primitive legs and other features suited for life out of the water. But paleontologists had not found fossils of the transitional animals between shallow-water fishes and limbed animals.

The team that discovered the new fossil decided to focus on far northern Canada when they noticed in a textbook that the region contained sedimentary rock deposited about 375 million years ago, just when shallow-water fishes were predicted by evolutionary science to be making the transition to land. The team had to travel for hours in planes and helicopters to reach the site, and they could work for just a couple of months each summer before snow began to fall. In their fourth summer of fieldwork they found what they had predicted they would find. In an outcropping of rock on the side of a hill, they uncovered the fossil of a creature that they named *Tiktaalik*. (The name means "big freshwater fish" in the language of the Inuit of northern Canada.) *Tiktaalik* still had many

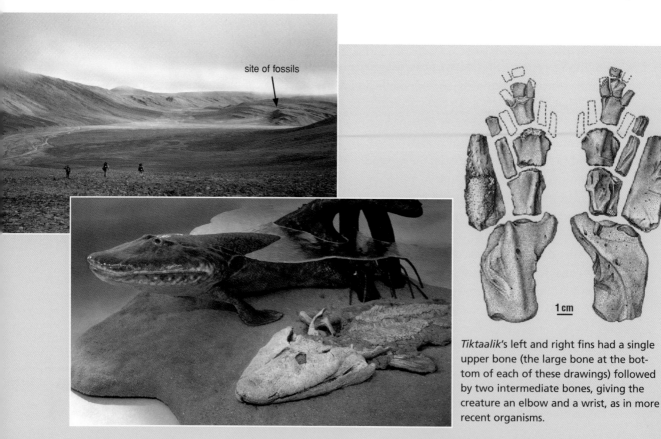

site of fossils

1 cm

*Tiktaalik*'s left and right fins had a single upper bone (the large bone at the bottom of each of these drawings) followed by two intermediate bones, giving the creature an elbow and a wrist, as in more recent organisms.

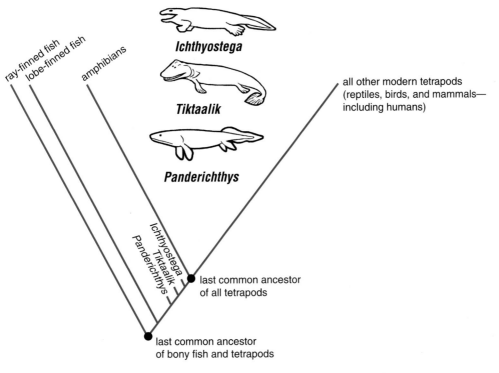

last common ancestor of all tetrapods

last common ancestor of bony fish and tetrapods

ray-finned fish

lobe-finned fish

amphibians

Ichthyostega

Tiktaalik

Panderichthys

all other modern tetrapods (reptiles, birds, and mammals—including humans)

*Tiktaalik* lived during the period when freshwater fishes were evolving the adaptations that enabled four-legged animals to live out of water. *Tiktaalik* may have lived somewhat before or somewhat after the ancestral species that gave rise to all of today's limbed animals, including humans. The evolutionary lineage that contained *Tiktaalik* may have gone extinct, as shown in this diagram by the short line branching from the main evolutionary lineage, or it may have been part of the evolutionary line leading to all modern tetrapods (animals with four legs). The last common ancestor of humans and all modern fishes also gave rise to evolutionary lineages that led to modern lobe-finned fishes (represented today by the coelacanth). In this and succeeding figures, time is represented by the lengths of the lines; modern groups of organisms are listed at the top of the figure.

of the features of fish, but it also had traits characteristic of early tetrapods. Most important, its fins contained bones that formed a limb-like appendage that the animal could use to move and prop itself up.

A prediction from more than a century of findings from evolutionary biology suggests that one of the early species that emerged from the Earth's oceans about 375 million years ago was the ancestor of amphibians, reptiles, dinosaurs, birds, and mammals. The discovery of *Tiktaalik* strongly supports that prediction. Indeed, the major bones in our own arms and legs are similar in overall configuration to those of *Tiktaalik*.

The discovery of *Tiktaalik*, while critically important for confirming predictions of evolution theory, is just one example of the many findings made every year that add depth and breadth to the scientific understanding of biological evolution. These discoveries come not just from paleontology but also from physics, chemistry, astronomy, and fields within biology. The theory of evolution is supported by so many observations and experiments that the overwhelming majority of scientists no longer question whether evolution has occurred and continues to occur and instead investigate the processes of evolution. Scientists are confident that the basic components of evolution will continue to be supported by new evidence, as they have been for the past 150 years.

# Biological evolution is the central organizing principle of modern biology.

The study of biological evolution has transformed our understanding of life on this planet. Evolution provides a scientific explanation for why there are so many different kinds of organisms on Earth and how all organisms on this planet are part of an evolutionary lineage. It demonstrates why some organisms that look quite different are in fact related, while other organisms that may look similar are only distantly related. It accounts for the appearance of humans on Earth and reveals our species' biological connections with other living things. It details how different groups of humans are related to each other and how we acquired many of our **traits.** It enables the development of effective new ways to protect ourselves against constantly evolving bacteria and viruses.

Biological evolution refers to changes in the traits of organisms over multiple generations. Until the development of the science of genetics at the beginning of the 20th century, biologists did not understand the mechanisms responsible for the inheritance of traits from parents to offspring. The study of genetics showed that heritable traits originate from the **DNA** that is passed from one generation to the next. DNA contains segments called genes that direct the production of **proteins** required for the growth and function of cells. Genes also orchestrate the development of a single-celled egg into a multicellular organism. DNA is therefore responsible for the continuity of biological form and function across generations.

However, offspring are not always exactly like their parents. Most organisms in any species, including humans, are genetically variable to some extent. In sexually reproducing species, where each parent contributes only one-half of its genetic information to its offspring (the offspring receives the full amount of genetic information when a sperm cell and an egg cell fuse), the DNA of the two parents is combined in new ways in the offspring. In addition, DNA can undergo changes known as **mutations** from one generation to the next, both in sexually reproducing and asexually reproducing organisms (such as bacteria).

When a mutation occurs in the DNA of an organism, several things can happen. The mutation may result in an altered trait that harms the organism, making it less likely to survive or produce offspring than other organisms in the **population** to which it belongs. Another possibility is that the mutation makes no difference to the well-being or reproductive success of an organism. Or the new mutation may result in a trait that enables an organism to take better advantage of the resources in its environment, thereby enhancing its ability to survive and produce offspring. For example, a fish might appear with a small modification to its fins that enables it to move more easily through shallow water (as occurred in the lineage leading to *Tiktaalik*); an insect might acquire a different shade of color that enables it to avoid being seen by predators; or a fly might have a difference in its wing patterns or courtship behaviors that more successfully attracts mates.

If a mutation increases the survivability of an organism, that organism is likely to have more offspring than other members of the population. If the offspring inherit the mutation, the number of organisms with the advantageous trait will increase from one generation to the next. In this way, the trait — and the genetic material (DNA) responsible for the trait — will tend to become more common in a population of organisms over time. In contrast, organisms possessing a harmful or deleterious mutation are less likely to contribute their DNA to future generations, and the trait resulting from the mutation will tend to become less frequent or will be eliminated in a population. Evolution consists of changes in the heritable traits of a population of organisms as successive generations replace one another. *It is populations of organisms that evolve, not individual organisms.*

The differential reproductive success of organisms with advantageous traits is known as **natural selection**, because nature "selects" traits that enhance the ability of organisms to survive and reproduce. Natural selection also can reduce the prevalence of traits that diminish organisms' abilities to survive and reproduce. Artificial selection is a similar process, but in this case humans rather than the environment select for desirable traits by arranging for animals or plants with those traits to breed. Artificial selection is the process responsible for the development of varieties of domestic animals (e.g., breeds of dogs, cats, and horses) and plants (e.g., roses, tulips, corn).

[**Natural selection:** *Differential survival and reproduction of organisms as a consequence of the characteristics of the environment.*]

## Evolution in Medicine: Combating New Infectious Diseases

In late 2002 several hundred people in China came down with a severe form of pneumonia caused by an unknown infectious agent. Dubbed "severe acute respiratory syndrome," or SARS, the disease soon spread to Vietnam, Hong Kong, and Canada and led to hundreds of deaths. In March 2003 a team of researchers at the University of California, San Francisco, received samples of a virus isolated from the tissues of a SARS patient. Using a new technology known as a DNA microarray, within 24 hours the researchers had identified the virus as a previously unknown member of a particular family of viruses — a result confirmed by other researchers using different techniques.

Immediately, work began on a blood test to identify people with the disease (so they could be quarantined), on treatments for the disease, and on vaccines to prevent infection with the virus.

An understanding of evolution was essential in the identification of the SARS virus. The genetic material in the virus was similar to that of other viruses because it had evolved from the same ancestor virus. Furthermore, knowledge of the evolutionary history of the SARS virus gave scientists important information about the disease, such as how it is spread. Knowing the evolutionary origins of human pathogens will be critical in the future as existing infectious agents evolve into new and more dangerous forms.

## Evolution in Agriculture: The Domestication of Wheat

When humans understand a phenomenon that occurs in nature, they often gain increased control over it or can adapt it to new uses. The domestication of wheat is a good example.

By recovering seeds from different archaeological sites and noticing changes in their characteristics over the centuries, scientists have hypothesized how wheat was altered by humans over time. About 11,000 years ago, people in the Middle East began growing plants for food rather than relying entirely on the wild plants and animals they could gather or hunt. These early farmers began saving seeds from plants with particularly favorable traits and planting those seeds in the next growing season. Through this process of "artificial selection," they created a variety of crops with characteristics particularly suited for agriculture. For example, farmers over many generations modified the traits of wild wheat so that seeds remained on the plant when ripe and could easily be separated from their hulls. Over the next few millennia, people around the world used similar processes of evolutionary change to transform many other wild plants and animals into the crops and domesticated animals we rely on today.

In recent years, plant scientists have begun making hybrids of wheat with some of their wild relatives from the Middle East and elsewhere. Using these hybrids, they have bred wheat varieties that are increasingly resistant to droughts, heat, and pests. Most recently, molecular biologists have been identifying the genes in the DNA of plants that are responsible for their advantageous traits so that these genes can be incorporated into other crops. These advances rely on an understanding of evolution to analyze the relationships among plants and to search for the traits that can be used to improve crops.

## *Evolution can result in both small and large changes in populations of organisms.*

Evolutionary biologists have discovered structures, biochemical processes and pathways, and behaviors that appear to have been highly conserved within and across species. Some species have undergone little overt change in their body structure over many millions of years. At the level of DNA, some genes that control the production of biochemicals or chemical reactions that are essential for cellular functioning show little variation across species that are only distantly related. (See, for example, the DNA sequences for two different genes that are conserved in closely related as well as more distantly related species that are described on pages 30 and 31.)

However, natural selection also can have radically different evolutionary effects over different timescales. Over periods of just a few generations (or, in some documented cases, even a single generation), evolution produces relatively small-scale **microevolutionary** changes in organisms. For example, many disease-causing bacteria have been evolving increased resistance to antibiotics. When a bacterium undergoes a genetic change that increases its ability to resist the effects of an antibiotic, that bacterium can survive and produce more copies of itself while nonresistant bacteria are being killed. Bacteria that cause tuberculosis, meningitis, staph infections, sexually transmitted diseases, and other illnesses have all become serious problems as they have developed resistance to an increasing number of antibiotics.

[**Microevolution:** *Changes in the traits of a group of organisms that do not result in a new species.*]

Another example of microevolutionary change comes from an experiment on the guppies that live in the Aripo River on the island of Trinidad. Guppies that live in the river are eaten by a larger species of fish that eats both juveniles and adults, while guppies that live in the small streams feeding into the river are eaten by a smaller fish that preys primarily on small juveniles. The guppies in the river mature faster, are smaller, and give birth to more and smaller offspring than the guppies in the streams do because guppies with these traits are better able to avoid their predator in the river than are larger guppies. When guppies were taken from the river and introduced into a stream without a preexisting population of guppies, they evolved traits like those of the stream guppies within about 20 generations.

Studies of guppies in Trinidad have demonstrated basic evolutionary mechanisms.

Incremental evolutionary changes can, over what are usually very long periods of time, give rise to new types of organisms, including new species. The formation of a new species generally occurs when one subgroup within a species mates for an extended period largely within the subgroup. For example, a subgroup may become geographically separated from the rest of the species, or a subgroup may come to use resources in a way that sets them apart from other members of the same species. As members of the subgroup mate among themselves, they accumulate genetic differences compared with the rest of the species. If this reproductive isolation continues for an extended period,

| How long could it take to produce 1,000 generations? How many generations might occur in a million years? | | | |
| --- | --- | --- | --- |
| | **1 Generation** | **1,000 Generations** | **Generations per 1 million years** |
| **Bacteria** | 1 hour to 1 day | 1,000 hours (42 days) to 2.7 years | 8.7 billion to 370.4 million |
| **Pets: dog/cat** | 2 years | 2,000 years | 500,000 |
| **Humans** | 22 years | 22,000 years | 45,000 |

When tetrapods (such as this sea turtle laying its eggs on a coastal beach) evolved the ability to lay hard-shelled eggs, they no longer had to return to the water to reproduce.

members of the subgroup may no longer respond to courtship or other signals from members of the original population. Eventually, genetic changes will become so substantial that the members of different subgroups can no longer produce viable offspring even if they do mate. In this way, existing species can continually "bud off" new species.

Over very long periods of time, continued instances of speciation can produce organisms that are very different from their ancestors. Though each new species resembles the species from which it arose, a succession of new species can diverge more and more from an ancestral form. This divergence from an ancestral form can be especially dramatic when an evolutionary change enables a group of organisms to occupy a new habitat or make use of resources in a novel way.

Consider, for example, the continued evolution of the tetrapods after limbed animals began living on land. As new species of plants evolved and covered the Earth, new species of tetrapods appeared with features that enabled them to take advantage of these new environments. The early tetrapods were amphibians that spent part of their lives on land but continued to lay their eggs in the water or in moist environments. The evolution about 340 million years ago of amniotic eggs, which have structures such as hard or leathery shells

The last common ancestor of the four-legged animals living today gave rise to amphibians and was the predecessor of reptiles. Birds and mammals evolved from different lineages of ancient reptiles.

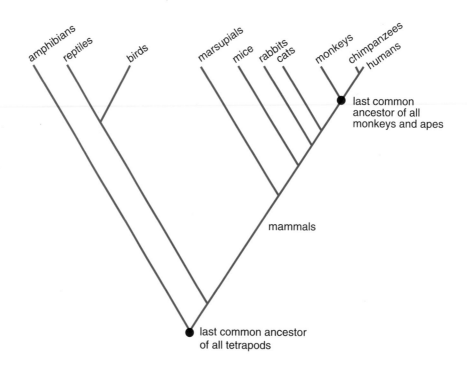

## Evolution in Industry: Putting Natural Selection to Work

The concept of natural selection has been applied in many fields outside biology. For example, chemists have applied principles of natural selection to develop new molecules with specific functions. First they create variants of an existing molecule using chemical techniques. They then test the variants for the desired function. The variants that do the best job are used to generate new variants. Repeated rounds of this selection process result in molecules that have a greatly enhanced ability to perform a given task. This technique has been used to create new enzymes that can convert cornstalks and other agricultural wastes into ethanol with increased efficiency.

and additional membranes that allow developing embryos to survive in dry environments, was one of the key developments in the evolution of the reptiles.

The early reptiles split into several major lineages. One lineage led to reptiles, including dinosaurs, and also to birds. Another lineage gave rise to mammals between 200 million and 250 million years ago.

The evolutionary transition from reptiles to mammals is particularly well documented in the fossil record. Successive fossil forms tend to have larger brains and more specialized sense organs, jaws and teeth adapted for more efficient chewing and eating, a gradual movement of the limbs from the sides of the body to under the body, and a female reproductive tract increasingly able to support the internal development and nourishment of young. Many of the biological novelties seen in mammals may be associated with the evolution of warm-bloodedness, which enabled a more active lifestyle over a much larger range of temperatures than in the cold-blooded reptilian ancestors.

Then, between 60 million and 80 million years ago, a group of mammals known as the primates first appeared in the fossil record. These mammals had grasping hands and feet, frontally directed eyes, and even larger and more complex brains. This is the lineage from which ancient and then modern humans evolved.

## Scientists seek explanations of natural phenomena based on empirical evidence.

Advances in the understanding of evolution over the past two centuries provide a superb example of how science works. Scientific knowledge and understanding accumulate from the interplay of observation and explanation. Scientists gather information by observing the natural world and conducting experiments. They then propose how the systems being studied behave in general, basing their explanations on the data provided through their experiments and other observations. They test their explanations by conducting additional observations and experiments under different conditions. Other scientists confirm the observations independently and carry out additional studies that may lead to more sophisticated explanations and predictions about future observations and experiments. In these ways, scientists continually arrive at more accurate and more comprehensive explanations of particular aspects of nature.

In science, explanations must be based on naturally occurring phenomena. Natural causes are, in principle, reproducible and therefore can be checked independently by others. If explanations are based on purported forces that are outside of nature, scientists have no way of either confirming or disproving those explanations. Any scientific explanation has to be *testable* — there must be possible observational consequences that could support the idea *but also ones that could refute it*. Unless a proposed explanation is framed in a way that some observational evidence could potentially count against it, that explanation cannot be subjected to scientific testing.

> ### *Definition of Science*
> **The use of evidence to construct testable explanations and predictions of natural phenomena, as well as the knowledge generated through this process.**

Because observations and explanations build on each other, science is a cumulative activity. Repeatable observations and experiments generate explanations that describe nature more accurately and comprehensively, and these explanations in turn suggest new observations and experiments that can be used to test and extend the explanation. In this way, the sophistication and scope of scientific explanations improve over time, as subsequent generations of scientists, often using technological innovations, work to correct, refine, and extend the work done by their predecessors.

# Is Evolution a Theory or a Fact?

It is both. But that answer requires looking more deeply at the meanings of the words "theory" and "fact."

In everyday usage, "theory" often refers to a hunch or a speculation. When people say, "I have a theory about why that happened," they are often drawing a conclusion based on fragmentary or inconclusive evidence.

The formal scientific definition of theory is quite different from the everyday meaning of the word. It refers to a comprehensive explanation of some aspect of nature that is supported by a vast body of evidence.

Many scientific theories are so well established that no new evidence is likely to alter them substantially. For example, no new evidence will demonstrate that the Earth does not orbit around the Sun (heliocentric theory), or that living things are not made of cells (cell theory), that matter is not composed of atoms, or that the surface of the Earth is not divided into solid plates that have moved over geological timescales (the theory of plate tectonics). Like these other foundational scientific theories, the theory of evolution is supported by so many observations and confirming experiments that scientists are confident that the basic components of the theory will not be overturned by new evidence. However, like all scientific theories, the theory of evolution is subject to continuing refinement as new areas of science emerge or as new technologies enable observations and experiments that were not possible previously.

One of the most useful properties of scientific theories is that they can be used to make predictions about natural events or phenomena that have not yet been observed. For example, the theory of gravitation predicted the behavior of objects on the Moon and other planets long before the activities of spacecraft and astronauts confirmed them. The evolutionary biologists who discovered *Tiktaalik* (see page 2) predicted that they would find fossils intermediate between fish and limbed terrestrial animals in sediments that were about 375 million years old. Their discovery confirmed the prediction made on the basis of evolutionary theory. In turn, confirmation of a prediction increases confidence in that theory.

In science, a "fact" typically refers to an observation, measurement, or other form of evidence that can be expected to occur the same way under similar circumstances. However, scientists also use the term "fact" to refer to a scientific explanation that has been tested and confirmed so many times that there is no longer a compelling reason to keep testing it or looking for additional examples. In that respect, the past and continuing occurrence of evolution is a scientific fact. Because the evidence supporting it is so strong, scientists no longer question whether biological evolution has occurred and is continuing to occur. Instead, they investigate the mechanisms of evolution, how rapidly evolution can take place, and related questions.

In science it is not possible to prove with absolute certainty that a given explanation is complete and final. Some of the explanations advanced by scientists turn out to be incorrect when they are tested by further observations or experiments. New instruments may make observations possible that reveal the inadequacy of an existing explanation. New ideas can lead to explanations that reveal the incompleteness or deficiencies of previous explanations. Many scientific ideas that once were accepted are now known to be inaccurate or to apply only within a limited domain.

However, many scientific explanations have been so thoroughly tested that they are very unlikely to change in substantial ways as new observations are made or new experiments are analyzed. These explanations are accepted by scientists as being true and factual descriptions of the natural world. The atomic structure of matter, the genetic basis of heredity, the circulation of blood, gravitation and planetary motion, and the process of biological evolution by natural selection are just a few examples of a very large number of scientific explanations that have been overwhelmingly substantiated.

Science is not the only way of knowing and understanding. *But science is a way of knowing that differs from other ways in its dependence on empirical evidence and testable explanations.* Because biological evolution accounts for events that are also central concerns of religion — including the origins of biological diversity and especially the origins of humans — evolution has been a contentious idea within society since it was first articulated by Charles Darwin and Alfred Russel Wallace in 1858.

# Acceptance of the evidence for evolution can be compatible with religious faith.

Today, many religious denominations accept that biological evolution has produced the diversity of living things over billions of years of Earth's history. Many have issued statements observing that evolution and the tenets of their faiths are compatible. Scientists and theologians have written eloquently about their awe and wonder at the history of the universe and of life on this planet, explaining that they see no conflict between their faith in God and the evidence for evolution. Religious denominations that do not accept the occurrence of evolution tend to be those that believe in strictly literal interpretations of religious texts.

Science and religion are based on different aspects of human experience. In science, explanations *must* be based on evidence drawn from examining the natural world. Scientifically based observations or experiments that conflict with an explanation eventually *must* lead to modification or even abandonment of that explanation. Religious faith, in contrast, does not depend only on empirical evidence, is not necessarily modified in the face of conflicting evidence, and typically involves supernatural forces or entities. Because they are not a part of nature, supernatural entities cannot be investigated by science. In this sense, science and religion are separate and address aspects of human understanding in different ways. Attempts to pit science and religion against each other create controversy where none needs to exist.

Many religious denominations and individual religious leaders have issued statements acknowledging the occurrence of evolution and pointing out that evolution and faith do not conflict.

"[T]here is no contradiction between an evolutionary theory of human origins and the doctrine of God as Creator."

— General Assembly of the Presbyterian Church

"[S]tudents' ignorance about evolution will seriously undermine their understanding of the world and the natural laws governing it, and their introduction to other explanations described as 'scientific' will give them false ideas about scientific methods and criteria."

— Central Conference of American Rabbis

"In his encyclical *Humani Generis* (1950), my predecessor Pius XII has already affirmed that there is no conflict between evolution and the doctrine of the faith regarding man and his vocation, provided that we do not lose sight of certain fixed points. . . . Today, more than a half-century after the appearance of that encyclical, some new findings lead us toward the recognition of evolution as more than an hypothesis. In fact it is remarkable that this theory has had progressively greater influence on the spirit of researchers, following a series of discoveries in different scholarly disciplines. The convergence in the results of these independent studies — which was neither planned nor sought — constitutes in itself a significant argument in favor of the theory."

— Pope John Paul II, Message to the Pontifical Academy of Sciences, October 22, 1996.

"We the undersigned, Christian clergy from many different traditions, believe that the timeless truths of the Bible and the discoveries of modern science may comfortably coexist. We believe that the theory of evolution is a foundational scientific truth, one that has stood up to rigorous scrutiny and upon which much of human knowledge and achievement rests. To reject this truth or to treat it as 'one theory among others' is to deliberately embrace scientific ignorance and transmit such ignorance to our children. We believe that among God's good gifts are human minds capable of critical thought and that the failure to fully employ this gift is a rejection of the will of our Creator. . . . We urge school board members to preserve the integrity of the science curriculum by affirming the teaching of the theory of evolution as a core component of human knowledge. We ask that science remain science and that religion remain religion, two very different, but complementary, forms of truth."

—"The Clergy Letter Project" signed by more than 10,000 Christian clergy members. For additional information, see http://www.butler.edu/clergyproject/clergy_project.htm.

# Excerpts of Statements by Scientists
## Who See No Conflict Between Their Faith and Science

Scientists, like people in other professions, hold a wide range of positions about religion and the role of supernatural forces or entities in the universe. Some adhere to a position known as scientism, which holds that the methods of science alone are sufficient for discovering everything there is to know about the universe. Others ascribe to an idea known as deism, which posits that God created all things and set the universe in motion but no longer actively directs physical phenomena. Others are theists, who believe that God actively intervenes in the world. Many scientists who believe in God, either as a prime mover or as an active force in the universe, have written eloquently about their beliefs.

"Creationists inevitably look for God in what science has not yet explained or in what they claim science cannot explain. Most scientists who are religious look for God in what science does understand and has explained."

— Kenneth Miller, professor of biology at Brown University and author of *Finding Darwin's God: A Scientist's Search for Common Ground Between God and Religion*. Quote is excerpted from an interview available at http://www.actionbioscience. org/evolution/miller.html.

"In my view, there is no conflict in being a rigorous scientist and a person who believes in a God who takes a personal interest in each one of us. Science's domain is to explore nature. God's domain is in the spiritual world, a realm not possible to explore with the tools and language of science. It must be examined with the heart, the mind, and the soul."

— Francis Collins, director of the Human Genome Project and of the National Human Genome Research Institute at the National Institutes of Health. Excerpted from his book, *The Language of God: A Scientist Presents Evidence for Belief* (p. 6).

"Our scientific understanding of the universe . . . provides for those who believe in God a marvelous opportunity to reflect upon their beliefs."

— Father George Coyne, Catholic priest and former director of the Vatican Observatory. Quote is from a talk, "Science Does Not Need God, or Does It? A Catholic Scientist Looks at Evolution," at Palm Beach Atlantic University, January 31, 2006. Available at http://chem.tufts. edu/AnswersInScience/Coyne-Evolution.htm.

# CHAPTER TWO

# THE EVIDENCE FOR BIOLOGICAL EVOLUTION

*Many areas of science have produced support for biological evolution.*

Many kinds of evidence have contributed to scientific understanding of biological evolution.  Some of this evidence — such as the fossils of long-extinct animals and the geographical distribution of species — was familiar to scientists in the 19th century or earlier.  Other forms of evidence — such as comparisons of DNA sequences — became available only in the 20th and 21st centuries.

The evidence for evolution comes not just from the biological sciences but also from both historical and modern research in anthropology, astrophysics, chemistry, geology, physics, mathematics, and other scientific disciplines, including the behavioral and social sciences.  Astrophysics and geology have demonstrated that the Earth is old enough for biological evolution to have resulted in the species seen today.  Physics and chemistry have led to dating methods that have established the timing of key evolutionary events.  Studies of other species have revealed not only the physical but also the behavioral continuities among species. Anthropology has provided new insights into human origins and the interactions between biology and cultural factors in shaping human behaviors and social systems.

As in every active area of science, many questions remain unanswered. Biologists continue to study the evolutionary relationships among organisms, the genetic changes that affect the form and function of organisms, the effects of organisms on Earth's physical environment, the evolution of intelligence and social behaviors, and many other fascinating subjects. But in each case they are asking specific questions to learn more about *how*, not *whether*, evolution has occurred and is continuing to occur. They are investigating and further elucidating the mechanisms that produce evolutionary change and the consequences of that change.

Biological evolution is part of a compelling historical narrative that scientists have constructed over the last few centuries. The narrative begins with the formation of the universe, the solar system, and the Earth, which resulted in the conditions necessary for life to evolve. While many questions remain about the origins of life on this planet, the appearance of life set in motion a process of biological evolution that continues to this day. Today, new chapters in the narrative are being uncovered through the study of the genetic processes responsible for evolutionary change.

## The origins of the universe, our galaxy, and our solar system produced the conditions necessary for the evolution of life on Earth.

The picture of Earth's place in the cosmos changed as much in the 20th century as it did in the 16th and 17th centuries following Copernicus's then controversial suggestion that the Sun, not the Earth, was at the center of the known universe. In the 1920s a new telescope at the Mount Wilson Observatory outside Los Angeles revealed that many of the faint smudges of light scattered across the night sky are not nebulae within our own Milky Way galaxy. Rather, they are separate galaxies, each containing many billions of stars. By studying the light emitted by these stars, astrophysicists arrived at another remarkable conclusion: The galaxies are receding from each other in every direction, which implies that the universe is expanding.

This observation led to the hypothesis first proposed by the Belgian astronomer and Roman Catholic priest Georges Lemaître that the universe originated in an event known as the "Big Bang." According to this idea, all of the energy and matter in the universe initially were compressed into an infinitesimally small, infinitely dense, and infinitely hot object known as a singularity, about which scientists still know very little. The universe then began to expand. As it did, the universe cooled to the point that the elementary particles that today form the matter of the universe became stable. The occurrence of the Big Bang, and the time that has elapsed since then, implied that matter in deep space should be at a particular temperature — a prediction confirmed by ground-

For ten consecutive days, the Hubble Space Telescope focused on a small patch of sky near the Big Dipper, revealing hundreds of galaxies never seen before.

based microwave radio telescopes. Later observations with satellites showed that the background radiation in the universe has exactly the properties that would be predicted from the Big Bang.

As the universe expanded, the matter in it gathered, by way of gravity and other processes that are not yet fully understood, into immense structures that became galaxies. Within these structures, much smaller clumps of matter collapsed into whirling clouds of gas and dust. When the matter in the center of an individual cloud became sufficiently compressed by gravity, the hydrogen atoms in that cloud began to fuse into helium atoms, giving off visible light and other radiation — the origin of a star.

Astrophysicists also have found that some stars form in the middle of a flattened spinning disk of matter. The gas and dust within such disks can aggregate into small grains, and these grains can form larger bodies called planetesimals. Computer simulations have indicated that planetesimals can coalesce into planets and other objects (such as moons and asteroids) orbiting a star. Our own solar system is likely to have formed in this way, and careful measurements have detected large planets orbiting stars in other parts of the Milky Way. These findings imply that billions of planets are orbiting the many billions of stars in our galaxy.

Astrophysicists and geologists have developed a variety of ways to measure the ages of the universe, our galaxy, the solar system, and the Earth. By measuring

A dark disk of dust and gas bisects a glowing star in this photograph from the Hubble Space Telescope. Such disks appear to provide the raw materials for the formation of planetesimals that combine to form planets and other orbiting bodies.

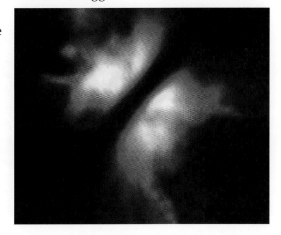

the distances between galaxies and the speeds with which they are separating, astronomers can calculate how much time has passed since the Big Bang. Increasingly accurate ways of measuring these quantities indicate that the universe is approximately 14 billion years old. Another way to estimate the universe's age, using measurements of the background radiation left behind by the Big Bang, produces similar results. Other observations and calculations suggest that our galaxy began to form a few hundred million years after the Big Bang, so the Milky Way is almost as old as the universe itself.

Our solar system formed within the Milky Way more recently. Measurements of radioactive elements in meteorites, which are the remnants of the materials that formed the solar system, indicate that our planet formed between 4.5 billion and 4.6 billion years ago. Asteroids and comets bombarded Earth after it formed, repeatedly melting the surface. Recent calculations show that one of

## Radiometric Dating

According to modern cosmology, the particles that constitute ordinary matter (protons, neutrons, and electrons) formed when the universe cooled after the Big Bang. These particles then came together to form hydrogen atoms, helium atoms, and small amounts of the next heavier element in the periodic table, lithium.

All the other elements in the universe were formed inside stars like the Sun and inside exploding stars known as supernovas. Through the addition of neutrons to lighter elements, nuclear reactions produced heavier elements. Supernovas dispersed these elements into interstellar space. Mixed with the hydrogen, helium, and lithium from the Big Bang, these elements formed our solar system.

Some atoms are radioactive, meaning that they naturally decay into other radioactive and nonradioactive atoms by emitting subatomic particles and energy. Each radioactive **nuclide** has a characteristic half-life, which is the amount of time it takes for half of the atoms in a sample to decay. Radioactive atoms therefore act as internal clocks for materials. By comparing the amount of a radioactive element in a material to the amount of its decay product, researchers can determine when the material formed. These measurements have yielded ages for the Earth, the Moon, meteorites, and the solar system. All of these measurements indicate that these objects are billions of years old.

Some who oppose the teaching of evolution try to cast doubt on radiometric age measurements. Radiometric dating is the product of more than a century of ingenious research and represents one of the most well-substantiated achievements of modern science.

**[Nuclide:** *An atom with a particular number of protons and neutrons in its nucleus. An element is defined by the number of protons in its nucleus. Nuclides that have the same number of protons but different numbers of neutrons are isotopes of that element.***]**

the objects to hit Earth was so large — about the size of Mars — that it splashed material into Earth's orbit that coalesced to form the Moon. The oldest rocks brought back from the Moon have ages measured to be 4.4 billion to 4.5 billion years. The oldest solid materials found on Earth are zircon crystals that formed 4.4 billion years ago. Rocks older than 3.5 billion years have been found on all the Earth's continents.

## *Living things appeared in the first billion years of Earth's history.*

Evidence from the most ancient fossils reveals that life has existed on Earth for most of our planet's history. Paleontologists working in Western Australia have discovered layered rocks known as stromatolites that appear to have resulted from the actions of bacteria at least 3.4 billion years ago, and fossils of cyanobacteria (also known as blue-green algae) have been determined to be nearly 3.5 billion years old. Other chemical evidence suggests that life may have originated much earlier, within a few hundred million years of when Earth's surface finally cooled.

Modern stromatolites formed by single-celled organisms (inset) closely resemble the structures formed by some of Earth's earliest living things.

Figuring out how life began is both an exciting and a challenging scientific problem. No fossil evidence of life forms older than 3.5 billion years has yet been found. Re-creating conditions that led to those earliest organisms is difficult because much remains unknown about the chemical and physical characteristics of the early Earth. Nevertheless, researchers have been developing hypotheses of how self-replicating organisms could form and begin to evolve, and they have tested the plausibility of these hypotheses in laboratories. While none of these hypotheses has yet achieved consensus, some progress has been made on these fundamental questions.

Since the 1950s hundreds of laboratory experiments have shown that Earth's simplest chemical compounds, including water and volcanic gases, could have reacted to form many of the molecular building blocks of life, including the molecules that make up proteins, DNA, and cell membranes. Meteorites from outer space also contain some of these chemical building blocks, and astronomers using radio telescopes have found many of these molecules in interstellar space.

For life to begin, three conditions had to be met. First, groups of molecules that could reproduce themselves had to come together. Second, copies of these molecular assemblages had to exhibit variation, so that some were better able to take advantage of resources and withstand challenges in the environment. Third, the variations had to be heritable, so that some variants would increase in number under favorable environmental conditions.

No one yet knows which combination of molecules first met these conditions, but researchers have shown how this process might have worked by studying a molecule known as **RNA**. Researchers recently discovered that some RNA molecules can greatly increase the rate of specific chemical reactions, including the replication of parts of other RNA molecules. If a molecule like RNA could reproduce itself (perhaps with the assistance of other molecules), it could form the basis for a very simple living organism. If such self-replicators were packaged within chemical vesicles or membranes, they might have formed "protocells"— early versions of very simple cells. Changes in these molecules could lead to variants that, for example, replicated more efficiently in a particular environment. In this way, natural selection would begin to operate, creating opportunities for protocells that had advantageous molecular innovations to increase in complexity.

Constructing a plausible hypothesis of life's origins will require that many questions be answered. Scientists who study the origin of life do not yet know which sets of chemicals could have begun replicating themselves. Even if a living cell could be made in the laboratory from simpler chemicals, it would not prove that nature followed the same pathway billions of years ago on the early Earth. But the principles underlying life's chemical origins, as well as plausible chemical details of the process, are subject to scientific investigation in the same ways that all other natural phenomena are. The history of science shows that even very difficult questions such as how life originated may become amenable to solution as a result of advances in theory, the development of new instrumentation, and the discovery of new facts.

[**RNA:** *Ribonucleic acid. A molecule related to DNA that consists of nucleotide subunits strung together in chains. RNA serves a number of cellular functions, including providing a template for the synthesis of proteins and catalyzing certain biochemical reactions.*]

## The fossil record provides extensive evidence documenting the occurrence of evolution.

[**Sedimentary:** *Rocks formed of particles deposited by water, wind, or ice.*]

Early in the 19th century, naturalists observed that fossils occur in a particular order in layers of **sedimentary** rock. Older materials are deposited more deeply and thus lie closer to the bottom of sedimentary rock than more recently deposited sediments, although older rocks can sometimes lie above younger rocks where large upheavals in the Earth's crust have taken place.

Fossils that closely resemble contemporary organisms appear in relatively young sediments, while fossils that only distantly resemble contemporary organisms occur in older sediments. Based on these observations, many naturalists, including Charles Darwin's grandfather, proposed that organisms had changed over time. But Darwin and Alfred Russel Wallace were the first to identify natural selection as the driving force behind evolution, or what Darwin termed "descent with modification."

When Darwin published *On the Origin of Species* in 1859, paleontology was still a rudimentary science. Sedimentary rocks from many time periods were unknown or had been inadequately studied. Darwin spent almost 20 years gathering evidence that supported his idea before making it public, but he also carefully considered evidential problems for his view, such as the inadequacy of the fossil record and the rarity of intermediate forms between some major groups of organisms at that time.

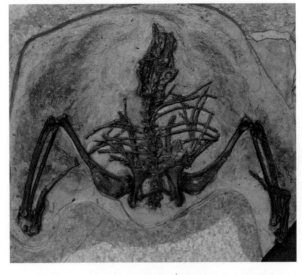

A near complete skeleton of a transitional bird-like fossil that was discovered in China and reported in 2006.

In the century and a half since then, paleontologists have found many intermediate forms that were not known in Darwin's time. In a variety of locations, sedimentary rocks that are between 540 million and 635 million years old contain traces of soft-bodied multicellular organisms, and fossilized tracks in earlier sediments hint at the existence of wormlike creatures as long ago as 1 billion years. Some of these organisms are likely to be the intermediate forms between the single-celled organisms that were Earth's sole inhabitants for the first 2 or more billion years of life's history and the hard-bodied organisms that appear in abundance in the fossil record beginning about 540 million years ago. Similarly, many of the organisms that appeared during this period were transitional forms between earlier soft-bodied organisms and major evolutionary lineages such as the fishes, arthropods, and mollusks that have survived to the present day.

As described at the beginning of this document, *Tiktaalik* is a notable transitional form between fish and the early tetrapods that lived on land. Fossils from about 330 million years ago document the evolution of large amphibians from the early tetrapods. Well-preserved skeletons from rocks that are 230 million years old show dinosaurs evolving from a lineage of reptiles. A long-standing example of a transitional form is *Archaeopteryx*, a 155-million-year-old fossil that has the skeleton of a small dinosaur but also feathers and wings. More birdlike fossils from China that are about 110 million years old have smaller tails and clawed appendages. In the more recent fossil record, the evolutionary paths of many modern organisms, such as whales, elephants, armadillos, horses, and humans, have been uncovered.

## *Common structures and behaviors often demonstrate that species have evolved from common ancestors.*

Each species that lives on Earth today is the product of an evolutionary lineage — that is, it arose from a preexisting species, which itself arose from a preexisting species, and so on back through time. For any two species living today, their evolutionary lineages can be traced back in time until the two lineages intersect. At that intersection is the species that was the most recent common ancestral species of the two modern species. (Sometimes, this common ancestral species is referred to as the common ancestor, but this term refers to a group of organisms rather than to a single ancestor.) For example, the common ancestor of humans and chimpanzees was a species estimated to have lived 6 to 7 million years ago, whereas the common ancestor of humans and the puffer fish was an ancient fish that lived in the Earth's oceans more than 400 million years ago.

*Thus, humans are not descended from chimpanzees or from any other ape living today but from a species that no longer exists.* Nor are humans descended from the species of fish that live today but, rather, from the species of fish that gave rise to the early tetrapods.

Modern chimpanzees, other great apes, and humans are descended from a common ancestor that is now extinct.

If the common ancestor of two species lived relatively recently, those two species are likely to have more physical features and behaviors in common than two species with a more distant common ancestor. Humans are thus far more similar to chimps than they are to fish. Nevertheless, all organisms share some common traits because they all share common ancestors at some point in the past. For example, based on accumulating fossil and molecular evidence, the common ancestor of humans, cows, whales, and bats was likely a small mammal that lived about 100 million years ago. The descendants of that common ancestor have undergone major changes, but their skeletons remain strikingly similar. A person writes, a cow walks, a whale swims, and a bat flies with structures built of bones that are different in detail but similar in general structure and relation to each other.

Evolutionary biologists call similar structures that derive from common ancestry "homologies." Comparative anatomists investigate such homologies, not only in bone structure but also in other parts of the body, and work out

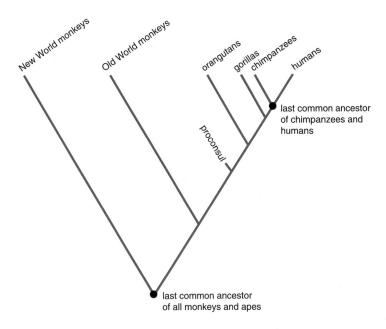

New World monkeys

Old World monkeys

orangutans

gorillas

chimpanzees

humans

proconsul

last common ancestor
of chimpanzees and
humans

last common ancestor
of all monkeys and apes

The last common ances-
tor of all monkeys and
apes lived about 40 mil-
lion years ago. Proconsul
was a species that lived
about 17 million years
ago. The most recent
species ancestral to both
humans and chimpan-
zees lived 6 to 7 million
years ago.

evolutionary relationships from degrees of similarity. Using the same logic, other biologists examine similarities in the functions of different organs, in the development of embryos, or in behaviors among different kinds of organisms. These investigations provide evidence about the evolutionary pathways that connect today's organisms to their common ancestors. Hypotheses based on this evidence then can be tested by examining the fossil record.

Sometimes, separate lineages independently evolve similar features, known as "analogous" structures, which look like homologies but result from common environments rather than common ancestry. For example, dolphins are aquatic mammals that have evolved from terrestrial mammals over the past 50 million

Though dolphins
(left) are more closely
related to humans
than they are to
sharks (right), they
have evolved bodies
adapted to an aquatic
environment. This is
an example of analo-
gous structures.

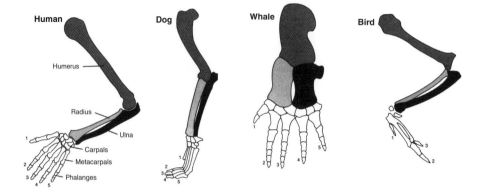

**Human**  **Dog**  **Whale**  **Bird**

Humerus

Radius

Ulna

Carpals

Metacarpals

Phalanges

The bones in the forelimbs of terrestrial and some aquatic vertebrates are remarkably similar because they have all evolved from the forelimbs of a common ancestor. This is an example of homologous structures.

years. In evolutionary terms, dolphins are as distant from fish as are mice or humans. But they have evolved streamlined bodies that closely resemble the bodies of fish, sharks, and even extinct dinosaurs known as ichthyosaurs. These kinds of evidence from many different fields of biology allow evolutionary biologists to discern whether physical and behavioral similarities are the product of common descent or are independent responses to similar environmental challenges.

## Evolution accounts for the geographic distribution of many plants and animals.

The diversity of life is almost unimaginable. Many millions of species live on, in, and above the Earth, each occupying its own ecological setting or niche. Some species, such as humans, dogs, and rats, can live in a wide range of environments. Others are extremely specialized. One species of a fungus grows exclusively on the rear portion of the covering on the wings of a single species of beetle that is found only in some caves in southern France. The larvae of the fly *Drosophila carcinophila* can develop only in specialized grooves beneath the flaps of the third pair of oral appendages of a land crab that is found solely on certain islands in the Caribbean.

The volcanic birth of the Hawaiian Islands in the Pacific Ocean over 2,000 miles from the nearest continent allowed one or a small number of windblown drosophilid flies such as the example pictured to evolve into more than 500 species in the islands' specialized environments. This rampant speciation was made possible in part because many of the environments in which they evolved were largely free of insect competitors and predators.

The occurrence of biological evolution both explains this diversity and accounts for its distribution. Consider, for example, the drosophilid flies of the Hawaiian Islands. More than 500 species of flies belonging to the genera *Drosophila* and the closely related *Scaptomyza* exist only in Hawaii. These Hawaiian species comprise about a quarter of all the species in these genera worldwide, and far more species than are found in a similar-sized area anywhere else on Earth. Why do so many different kinds of flies live exclusively in Hawaii?

The geological and biological history of Hawaii provides an answer. The Hawaiian Islands consist of the tops of mid-ocean volcanoes and have never been connected to any body of land. The islands formed as the Pacific tectonic plate moved over a "hot spot" where upwelling molten rock from the Earth's interior heats the Earth's crust. The newest islands are the tallest, while older islands progressively erode and eventually sink beneath the water. Thus, the oldest landmass in the chain, Kure Atoll, rose from the Pacific about 30 million years ago, while the youngest, the "Big Island" of Hawaii, is only about 500,000 years old and still has considerable ongoing volcanic activity.

All of the native plants and animals of the Hawaiian Islands — that is, those existing on the islands before the arrival of humans 1,200 to 1,600 years ago — are descended from organisms that made their way through the air or the water from the surrounding continents and from distant islands to the initially barren islands. In the case of the Hawaiian drosophilids, several lines of evidence, especially from DNA, indicate that all of the native *Drosophila* and *Scaptomyza* species are descended from a single ancestral species that colonized the islands millions of years ago.

These initial colonizers encountered conditions that were very favorable to rapid **speciation**. Individual species repeatedly served as ancestors for multiple other species as groups of flies occupied habitats with different elevations, precipitation, soils, and plants. In addition, small groups of flies — or in some cases perhaps a single fertilized female — periodically flew or were carried to other islands, where they gave rise to new species. Many new species were able to occupy ecological niches that on the continents already would have been filled by other species. For example, many Hawaiian drosophilids lay eggs in decaying leaves on the ground, an ecological niche that is filled by insects and other organisms on the continents but in the Hawaiian Islands was almost empty.

The mammals that have lived in North and South America provide another good example of how evolution accounts for the distribution of species. These two continents were connected as part of a much larger landmass during the early evolution of the mammals. But the breakup of that landmass caused North and South America to separate, after which their respective mammals evolved in different directions. The mammals that evolved in South America include such modern-day groups as anteaters, sloths, opossums, and armadillos, according to the fossil record. In North America, horses, bats, wolves, and

[**Speciation:** *The evolutionary processes through which new species arise from existing species.*]

When tectonic forces joined North and South America, mammals that had evolved in South America, such as the armadillo, migrated north.

the saber-toothed cat were among the many species that evolved. Then, about 3 million years ago, North and South America were reconnected as a consequence of the movement of the Earth's tectonic plates. Mammals of South American origin, such as armadillos, porcupines, and opossums, migrated north. Meanwhile, many kinds of North American mammals, including deer, raccoons, mountain lions, bears, and dogs, eventually made their way across the isthmus to the south.

## Molecular biology has confirmed and extended the conclusions about evolution drawn from other forms of evidence.

Charles Darwin and other 19th-century biologists arrived at their conclusions despite knowing almost nothing about the molecular basis of life. Since then, the ability to examine biological molecules in detail has provided an entirely new form of evidence about the mechanisms and historical pathways of evolution. This new evidence has fully confirmed the general conclusions drawn from the fossil record, the geographic distribution of species, and other types of observations. In addition, it has provided a wealth of new information about the evolutionary relationships among species and about how evolution occurs.

DNA is passed from one generation to the next directly from a parent to its offspring (in asexually reproducing organisms) or through the union of DNA-containing sperm and egg cells (in sexually reproducing organisms). As discussed earlier, the sequence of nucleotides in DNA can change from one generation to the next because of mutations; if these changes give rise to beneficial traits, the new DNA sequences are likely to spread within a population

# The Picture-Winged Drosophilids

The drosophilid flies of Hawaii provide an excellent example of "adaptive radiation," in which an ancestral species gives rise to a very large number of new species in a relatively short time. Evolutionary biologists have focused particular attention on a group of about 100 drosophilid species that have characteristic pigmented markings on their large wings. Known as the picture-winged drosophilids, these species carry within them a remarkable biological record of the group's evolutionary history.

[**Chromosome:** *A double stranded DNA molecule that contains a series of specific genes along its length. In most sexually reproducing organisms, chromosomes occur in pairs, with one member of the pair being inherited from each parent.*]

Cells in the salivary glands of all *Drosophila* larvae contain special chromosomal structures known as polytene chromosomes. Easily visible through a microscope, these polytene chromosomes display hundreds of alternating dark and light bands of different sizes. These banding patterns make it especially easy to detect a kind of chromosomal rearrangement known as an inversion. Sometimes, a mistake during the duplication of DNA can cause a segment of the chromosome to be flipped. The result is a rearranged chromosome in which a section of the chromosome, with its characteristic light and dark bands, has a reversed orientation. Many inversions of this type have occurred in different segments of chromosomes in different species of flies.

As individual species of drosophilids on the Hawaiian islands have diversified to form multiple species, researchers have used the resulting changes in banding patterns to reconstruct the sequence in which existing species of drosophilids moved from older islands to newer islands and gave rise to new species. For example, the "Big Island" of Hawaii, which is the youngest in the island chain, currently has 26 species of picture-winged drosophilids. By examining the specific chromosome inversions in these colonizing species and comparing them with species that live on islands that are older, researchers have determined that flies on the Big Island have all originated from 19 separate colonizations of the island by a small group of flies (or perhaps single fertilized female flies) from one of the older islands.

*Photograph of a polytene chromosome from a* Drosophila *larva shows two breakpoints (indicated by solid bars) where a portion of the chromosome is inverted compared to the same chromosome in other species.*

over multiple generations. In addition, neutral mutations that have no effect on the traits of an organism can be maintained within a population as DNA passes between generations. As a result, DNA contains a record of past genetic changes, including the changes responsible for evolutionary adaptations.

By comparing the DNA sequences of two organisms, biologists can uncover the genetic changes that have occurred since those organisms shared a common ancestor. If two species have a relatively recent common ancestor, their DNA sequences will be more similar than the DNA sequences for two species that share a distant common ancestor. For example, the DNA sequences of humans, which vary to a small degree among individuals and populations

The gene that, when mutated, causes cystic fibrosis in humans is very similar to the corresponding gene in chimpanzees but is less similar to the corresponding gene in organisms that are less closely related to humans. The height of the green bars shows the similarity of the gene in other organisms to the human gene over a span of 10,000 nucleotides.

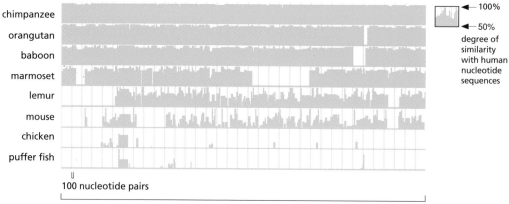

chimpanzee
orangutan
baboon
marmoset
lemur
mouse
chicken
puffer fish

100 nucleotide pairs

10,000 nucleotide pairs

100%
50%
degree of similarity with human nucleotide sequences

of people, on average differ by just a few percent from those of chimpanzees, reflecting our relatively recent common ancestry. But human DNA sequences are increasingly different from those of the baboon, mouse, chicken, and puffer fish, reflecting our increasing evolutionary distance from each of those organisms. Even greater differences in DNA sequences are found when comparing humans to flies, worms, and plants. Yet similarities in DNA sequences can be seen across all living forms, despite the amount of time that has elapsed since

## The Evolution of Limbs in Early Tetrapods

Molecular biologists have been discovering DNA regions that control the formation of body parts during development. Some of the most important of these DNA regions are known as *Hox* genes.

Humans and all other mammals have 39 *Hox* genes. Individual *Hox* genes control the function of other types of genes, and the same *Hox* gene can control different sets of genes in different parts of the body.

*Hox* genes are also involved in the development of many different anatomical features, including limbs, the spine, the digestive system, and the reproductive tract in diverse species of both invertebrate and vertebrate animals. For example, as illustrated in the figure (right side of page), the same *Hox* genes that control the development of body parts in the fruit fly *Drosophila* also control the development of body parts in mice and other mammals. Colors indicate the activity of the same *Hox* gene in both kinds of organisms.

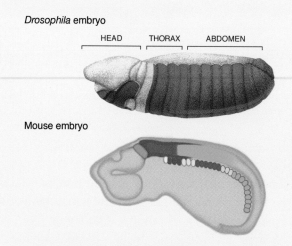

*Drosophila* embryo

HEAD    THORAX    ABDOMEN

Mouse embryo

*Hox* genes also direct the formation of fins in fish and limbs in land-dwelling vertebrates. They are expressed in different patterns in limbed animals, resulting in the formation of fingers and toes. Changes in the expression of these genes were likely involved in the evolution of the early tetrapods, such as *Tiktaalik*.

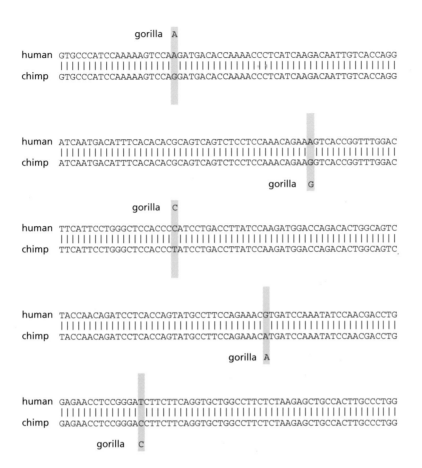

Comparison of the human and chimp DNA sequences for the gene that encodes the hormone leptin (which is involved in the metabolism of fats) reveals only five differences in 250 nucleotides. Where the human and chimpanzee sequences differ, the corresponding nucleotide in the gorilla (shaded bars) can be used to derive the nucleotide that likely existed in the common ancestor of humans, chimpanzees, and gorillas. In two cases, the gorilla and human nucleotides match, while in the other three cases, the gorilla and chimpanzee sequences are the same. The common ancestor of the gorilla, chimpanzee, and human is most likely to have had the nucleotide that is the same in two of the three modern-day organisms because this would require just one DNA change rather than two.

they had common ancestors. Even humans and bacteria share some similarity in DNA sequences in certain genes, and these similarities correspond to molecular systems with similar functions. Biological evolution thus explains why other organisms can be studied to understand biological processes critical to human life. Indeed, much of the biomedical research carried out today is based on the biological commonalities of all living things.

The study of biological molecules has done more than document the evolutionary relationships among organisms. It also can reveal how genetic changes produce new traits in organisms over the course of evolutionary history. For example, molecular biologists have been examining the function of regulatory proteins that cause other genes in a cell to turn on and off as an organism develops from a fertilized egg. Small changes in these proteins, in the DNA regions to which these proteins attach, or even, as recently discovered, in small RNA molecules can have dramatic effects on the anatomy and function of an organism. Such changes could be responsible for some of the major evolutionary innovations that have occurred over time, such as the development of limbs from fins in early tetrapods. Moreover, biologists have discovered that very similar sets of regulatory proteins occur in organisms as different as flies, mice, and humans, despite the many millions of years that separate these organisms from their common ancestors. The DNA evidence suggests that the basic mechanisms controlling biological form became established before or during the evolution of multicellular organisms and have been conserved with little modification ever since.

## *Biological evolution explains the origin and history of our species.*

Study of all the forms of evidence discussed earlier in this booklet has led to the conclusion that humans evolved from ancestral primates. In the 19th century, the idea that humans and apes had common ancestors was a novel one, and it was hotly debated among scientists in Darwin's time and for years after.

## The Evolution of Whales, Dolphins, and Porpoises

The combination of fossil and molecular evidence enables biologists to construct much more detailed evolutionary histories than have been possible in the past. For example, recent fossil discoveries in

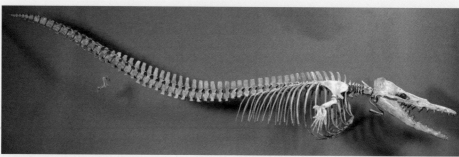

Asia have revealed a succession of organisms that, beginning about 50 million years ago, moved from life on land first to hunt and then to live continuously in marine environments. This fossil evidence accords with recent genetic findings that whales, dolphins, and porpoises are descended from a group of terrestrial mammals known as artiodactyls, which today includes such animals as sheep, goats, and giraffes. Most recently, studies of regulatory networks in the DNA of modern porpoises have revealed the molecular changes that caused the ancestors of these organisms to lose their hind limbs and develop more streamlined bodies. All of these forms of evidence support each other and add fascinating details to the understanding of evolution.

*Fossils of* Dorudon, *found in Egypt and dating to approximately 40 million years ago, document a critical transition in the evolution of modern whales. Because it had evolved from a mammal that lived on land,* Dorudon *still had vestigial traces of hind limbs, feet, and toes (the small bones at the base of the tail), even though it lived in the water and used its long powerful tail to swim.*

More than 3.5 million years ago, two hominids walked upright across a field of newly fallen volcanic ash in eastern Africa. The footprints were covered by a subsequent ashfall until 1978, when they were unearthed by paleontologists. The Laetoli footprints, named after the site where they were found, are very early evidence of upright walking, a key acquisition in the lineage leading to humans.

But today there is no scientific doubt about the close evolutionary relationships between humans and all other primates. Using the same scientific methods and tools that have been employed to study the evolution of other species, researchers have compiled a large and increasing number of fossil discoveries and compelling new molecular evidence that clearly indicate that the same forces responsible for the evolution of all other life forms on Earth account for the biological evolution of human characteristics.

Based on the strength of evidence from DNA comparisons, the common ancestor of humans and chimpanzees lived approximately 6 to 7 million years ago in Africa. The evolutionary tree leading from this ancestral species to modern humans contains a number of side branches, representing populations and species that eventually went extinct. At various times in the past, the planet appears to have been populated by several human-like species.

About 4.1 million years ago, a species appeared in Africa that paleontologists place in the genus *Australopithecus,* which means "southern ape." (A member of the genus was first discovered in southern Africa, although other fossils, including an almost complete skeleton of a 3-year-old female, have been found in eastern Africa.) The brain of an adult of this genus was about the same size as that of modern apes (as documented by the size of fossil

In the drawing at right, the skeleton of Lucy, exemplar of an adult member of the species *Australopithecus afarensis* (with shaded bones representing those that were recovered), dates from the same geological period when the Laetoli footprints were made. For comparison, the skeleton of a modern human stands beside her.

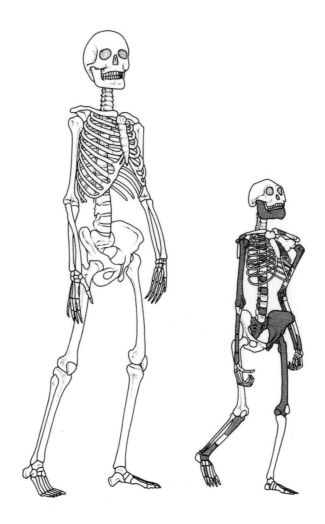

skulls), and it appears to have spent part of its life climbing in trees, as indicated by its short legs and features of its upper limbs. But *Australopithecus* also walked upright, as humans do. Footprints left by one of the earliest *Australopithecus* species have been discovered preserved with remarkable clarity in hardened volcanic ash.

About 2.3 million years ago, the earliest species of *Homo*, the genus to which all modern humans belong, evolved in Africa. This species is known as *Homo habilis* ("handy" or "skillful man"). Its average brain size, as determined from skulls that postdate 2 million years ago, was probably about 50 percent larger than that of earlier *Australopithecus*. The earliest stone tools appear about 2.6 million years ago.

About 1.8 million years ago, a more evolved species, *Homo erectus* ("upright man") appeared. This species spread from Africa to Eurasia. The subsequent fossil record includes the skeletal remains of additional species within the genus *Homo*. The more recent species generally had larger brains than the earlier ones.

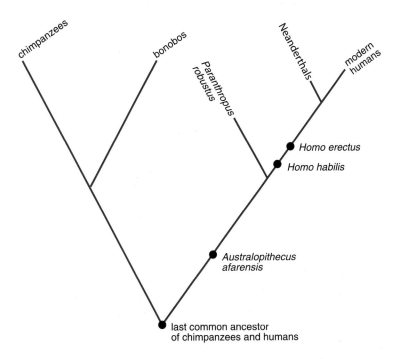

chimpanzees

bonobos

*Paranthropus robustus*

Neanderthals

modern humans

● *Homo erectus*

● *Homo habilis*

● *Australopithecus afarensis*

● last common ancestor of chimpanzees and humans

A number of species, of which only *Australopithecus afarensis*, *Homo habilis*, and *Homo erectus* are shown here, are thought to represent evolutionary links between modern humans and the more ancient species that was the common ancestor of chimpanzees, bonobos (a close relative of chimpanzees), and modern humans. Other closely related species on the human side of the family tree are known from the fossil record. *Paranthropus robustus* and Neanderthals are extinct evolutionary lineages now represented only by fossils.

Evidence shows that anatomically modern humans (*Homo sapiens* —"wise" or "knowing man") with bodies and brains like ours, evolved in Africa from earlier forms of humans. The earliest known fossil of a modern human is less than 200,000 years old. The members of this group dispersed throughout Africa and, more recently, into Asia, Australia, Europe, and the Americas, replacing earlier populations of humans then living in some parts of the world. ■

# CHAPTER THREE

# CREATIONIST PERSPECTIVES

## *Creationist views reject scientific findings and methods.*

Advocates of the ideas collectively known as "creationism" and, recently, "intelligent design creationism" hold a wide variety of views. Most broadly, a "creationist" is someone who rejects natural scientific explanations of the known universe in favor of special creation by a supernatural entity. Creationism in its various forms is not the same thing as belief in God because, as was discussed earlier, many believers as well as many mainstream religious groups accept the findings of science, including evolution. Nor is creationism necessarily tied to Christians who interpret the Bible literally. Some non-Christian religious believers also want to replace scientific explanations with their own religion's supernatural accounts of physical phenomena.

In the United States, various views of creationism typically have been promoted by small groups of politically active religious fundamentalists who believe that only a supernatural entity could account for the physical changes in the universe and for the biological diversity of life on Earth. But even these creationists hold very different views. Some, known as "young Earth" creationists, believe the biblical account that the universe and the Earth were created just a few thousand years ago. Proponents of this form of creationism also believe that all living things, including humans, were created in a very short period of time in essentially the forms in which they exist today. Other creationists, known as "old Earth" creationists, accept that the Earth may be very old but reject other scientific findings regarding the evolution of living things.

No scientific evidence supports these viewpoints. On the contrary, as discussed earlier, several independent lines of evidence indicate that the Earth is about 4.5 billion years old and that the universe is about 14 billion years old. Rejecting the evidence for these age estimates would mean rejecting not just biological evolution but also fundamental discoveries of modern physics, chemistry, astrophysics, and geology.

Some creationists believe that Earth's present form and the distribution of fossils can be explained by a worldwide flood. But this claim also is at odds with observations and evidence understood scientifically. The belief that Earth's sediments, with their fossils, were deposited in a short period does not accord either with the known processes of sedimentation or with the estimated volume of water needed to deposit sediments on the top of some of Earth's highest mountains.

Creationists sometimes cite what they claim to be an incomplete fossil record as evidence that living things were created in their modern forms. But this argument ignores the rich and extremely detailed record of evolutionary history that paleontologists and other biologists have constructed over the past two centuries and are continuing to construct. Paleontological research has filled in many of the parts of the fossil record that were incomplete in Charles Darwin's time. The claim that the fossil record is "full of gaps" that undermine evolution is simply false. Indeed, paleontologists now know enough about the ages of sediments to predict where they will be able to find particularly significant transitional fossils, as happened with *Tiktaalik* and the ancestors of modern humans. Researchers also are using new techniques, such as computed axial tomography (**CT**), to learn even more about the internal structures and composition of delicate bones of fossils. Exciting new discoveries of fossils continue to be reported in both the scientific literature and popular media.

Another compelling feature of the fossil record is its consistency. Nowhere on Earth are fossils from dinosaurs, which went extinct 65 million years ago, found together with fossils from humans, who evolved in just the last few million years. Nowhere are the fossils of mammals found in sediments that are more than about 220 million years old. Whenever creationists point to sediments where these relationships appear to be altered or even reversed, scientists have clearly demonstrated that this reversal has resulted from the folding of geological strata over or under others. Sediments containing the fossils of only unicellular organisms appear earlier in the fossil record than do sediments containing the remains of both unicellular and multicellular organisms. The sequence of fossils across Earth's sediments points unambiguously toward the occurrence of evolution.

[**CT:** *A medical imaging technique that generates a three-dimensional view of some object by combining a series of two-dimensional X-ray images of "slices" of that object.*]

Creationists sometimes argue that the idea of evolution must remain hypothetical because "no one has ever seen evolution occur." This kind of statement also reveals that some creationists misunderstand an important characteristic of scientific reasoning. Scientific conclusions are not limited to direct observation but often depend on inferences that are made by applying reason to observations. Even with the launch of Earth-orbiting spacecraft, scientists could not directly see the Earth going around the Sun. But they inferred from a wealth of independent measurements that the Sun is at the center of the solar system. Until the recent development of extremely powerful microscopes, scientists could not observe atoms, but the behavior of physical objects left no doubt about the atomic nature of matter. Scientists hypothesized the existence of viruses for many years before microscopes became powerful enough to see them.

Thus, for many areas of science, scientists have not directly observed the objects (such as genes and atoms) or the phenomena (such as the Earth going around the Sun) that are now well-established facts. Instead, they have confirmed them indirectly by observational and experimental evidence. Evolution is no different. Indeed, for the reasons described in this booklet, evolutionary science provides one of the best examples of a deep understanding based on scientific reasoning.

This contention that nobody has seen evolution occurring further ignores the overwhelming evidence that evolution has taken place and is continuing to occur. The annual changes in influenza viruses and the emergence of bacteria resistant to antibiotics are both products of evolutionary forces. Another example of ongoing evolution is the appearance of mosquitoes resistant to various insecticides, which has contributed to a resurgence of malaria in Africa and elsewhere. The transitional fossils that have been found in abundance since Darwin's time reveal how species continually give rise to successor species that, over time, produce radically changed body forms and functions. It also is possible to directly observe many of the specific processes by which evolution occurs. Scientists regularly do experiments using microbes and other model systems that directly test evolutionary hypotheses.

Creationists reject such scientific facts in part because they do not accept evidence drawn from natural processes that they consider to be at odds with the Bible. But science cannot test supernatural possibilities. To young Earth creationists, no amount of empirical evidence that the Earth is billions of years old is likely to refute their claim that the world is actually young but that God simply made it *appear* to be old. Because such appeals to the supernatural are not testable using the rules and processes of scientific inquiry, they cannot be a part of science.

## "Intelligent design" creationism is not supported by scientific evidence.

Some members of a newer school of creationists have temporarily set aside the question of whether the solar system, the galaxy, and the universe are billions or just thousands of years old. But these creationists unite in contending that the physical universe and living things show evidence of "intelligent design." They argue that certain biological structures are so complex that they could not have evolved through processes of undirected mutation and natural selection, a condition they call "irreducible complexity." Echoing theological arguments that predate the theory of evolution, they contend that biological organisms must be designed in the same way that a mousetrap or a clock is designed — that in order for the device to work properly, all of its components must be available simultaneously. If one component is missing or changed, the device will fail to operate properly. Because even such "simple" biological structures as the flagellum of a bacterium are so complex, proponents of intelligent design creationism argue that the probability of all of their components being produced and simultaneously available through random processes of mutation are infinitesimally small. The appearance of more complex biological structures (such as the vertebrate eye) or functions (such as the immune system) is impossible through natural processes, according to this view, and so must be attributed to a transcendent intelligent designer.

Electron micrograph of a bacterium with hair-like flagella.

However, the claims of intelligent design creationists are disproven by the findings of modern biology. Biologists have examined each of the molecular systems claimed to be the products of design and have shown how they could have arisen through natural processes. For example, in the case of the bacterial flagellum, there is no single, uniform structure that is found in all flagellar bacteria. There are many types of flagella, some simpler than others, and many species of bacteria do not have flagella to aid in their movement. Thus, other components of bacterial cell membranes are likely the precursors of the proteins found in various flagella. In addition, some bacteria inject toxins into other cells through proteins that are secreted from the bacterium and that are very similar in their molecular structure to the proteins in parts of flagella. This similarity indicates a common evolutionary origin, where small changes in the structure and organization of secretory proteins could serve as the basis for flagellar proteins. Thus, flagellar proteins are not irreducibly complex.

Evolutionary biologists also have demonstrated how complex biochemical mechanisms, such as the clotting of blood or the mammalian immune system, could have evolved from simpler precursor systems. With the clotting of blood, some of the components of the mammalian system were present in earlier organisms, as demonstrated by the organisms living today (such as fish, reptiles, and birds) that are descended from these mammalian precursors. Mammalian clotting systems have built on these earlier components.

Existing systems also can acquire new functions. For example, a particular system might have one task in a cell and then become adapted through evolutionary processes for different use. The *Hox* genes (described in the box on page 30) are a prime example of evolution finding new uses for existing systems. Molecular biologists have discovered that a particularly important mechanism through which biological systems acquire additional functions is gene duplication. Segments of DNA are frequently duplicated when cells divide, so that a cell has multiple copies of one or more genes. If these multiple copies are passed on to offspring, one copy of a gene can serve the original function in a cell while the other copy is able to accumulate changes that ultimately result in a new function. The biochemical mechanisms responsible for many cellular processes show clear evidence for historical duplications of DNA regions.

In addition to its scientific failings, this and other standard creationist arguments are fallacious in that they are based on a false dichotomy. Even if their negative arguments against evolution were correct, that would not establish the creationists' claims. There may be alternative explanations. For example, it would be incorrect to conclude that because there is no evidence that it is raining outside, it must be sunny. Other explanations also might be possible. Science requires testable evidence for a hypothesis, not just challenges against

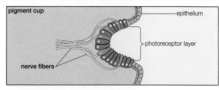
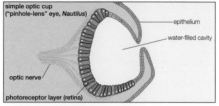
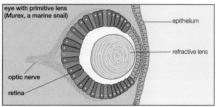
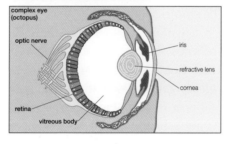

Eyes in living mollusks. The octopus eye (bottom) is quite complex, with components similar to those of the human eye, such as a cornea, iris, refractive lens, and retina. Other mollusks have simpler eyes. The simplest eye is found in limpets (top), consisting of only a few pigmented cells, slightly modified from typical epithelial (skin) cells. Slit-shell mollusks (second from top) have a slightly more advanced organ, consisting of some pigmented cells shaped as a cup. Further elaborations and increasing complexity are found in the eyes of *Nautilus* and *Murex*, which are not as complex as the eyes of the squid and octopus.

Over millions of years, the Colorado River has cut through the rocks of the Colorado plateau, revealing sedimentary rocks deposited more than a billion years ago.

one's opponent. Intelligent design is not a scientific concept because it cannot be empirically tested.

Creationists sometimes claim that scientists have a vested interest in the concept of biological evolution and are unwilling to consider other possibilities. But this claim, too, misrepresents science. Scientists continually test their ideas against observations and submit their work to their colleagues for critical peer review of ideas, evidence, and conclusions before a scientific paper is published in any respected scientific journal. Unexplained observations are eagerly pursued because they can be signs of important new science or problems with an existing hypothesis or theory. History is replete with scientists challenging accepted theory by offering new evidence and more comprehensive explanations to account for natural phenomena. Also, science has a competitive element as well as a cooperative one. If one scientist clings to particular ideas despite evidence to the contrary, another scientist will attempt to replicate relevant experiments and will not hesitate to publish conflicting evidence. If there were serious problems in evolutionary science, many scientists would be eager to win fame by being the first to provide a better testable alternative. That there are no viable alternatives to evolution in the scientific literature is not because of vested interests or censorship but because evolution has been and continues to be solidly supported by evidence.

The potential utility of science also demands openness to new ideas. If petroleum geologists could find more oil and gas by interpreting the record of sedimentary rocks (where deposits of oil and natural gas are found) as having resulted from a single flood, they would certainly favor the idea of such a flood, but they do not. Instead, petroleum geologists agree with other geologists that sedimentary rocks are the products of billions of years of Earth's history. Indeed, petroleum geologists have been pioneers in the recognition of fossil deposits that were formed over millions of years in such environments as meandering rivers, deltas, sandy barrier beaches, and coral reefs.

The arguments of creationists reverse the scientific process. They begin with an explanation that they are unwilling to alter — that supernatural forces have shaped biological or Earth systems — rejecting the basic requirements of science that hypotheses must be restricted to testable natural explanations. Their beliefs cannot be tested, modified, or rejected by scientific means and thus cannot be a part of the processes of science.

## *The pressure to downplay evolution or emphasize nonscientific alternatives in public schools compromises science education.*

Despite the lack of scientific evidence for creationist positions, some advocates continue to demand that various forms of creationism be taught together with or in place of evolution in science classes. Many teachers are under considerable pressure from policy makers, school administrators, parents, and students to downplay or eliminate the teaching of evolution. As a result, many U.S. students lack access to information and ideas that are both integral to modern science and essential for making informed, evidence-based decisions about their own lives and our collective future.

Regardless of the careers that they ultimately select, to succeed in today's scientifically and technologically sophisticated world, *all* students need a sound education in science. Many of today's fast-growing and high-paying jobs require a familiarity with the core concepts, applications, and implications of science. To make informed decisions about public policies, people need to know how scientific evidence supports those policies and whether that evidence was gathered using well-established scientific practice and principles. Learning about evolution is an excellent way to help students understand the nature, processes, and limits of science in addition to concepts about this fundamentally important contribution to scientific knowledge.

Given the importance of science in all aspects of modern life, the science curriculum should not be undermined with nonscientific material. Teaching creationist ideas in science classes confuses what constitutes science and what does not. It compromises the objectives of public education and the goal of a high-quality science education.

# Excerpts from Court Cases

Since the 1925 trial of John Scopes, which investigated the legality of a Tennessee law that forbade the teaching in public schools of "any theory that denies the story of the Divine Creation of man as taught in the Bible," a number of court cases have looked at laws involving the teaching of creationist ideas. Several court decisions, including the 1987 Supreme Court case *Edwards v. Aguillard* and, more recently, the 2005 federal district court case (in central Pennsylvania) of *Kitzmiller v. Dover Area School District*, have ruled that the various forms of creationism, including intelligent design creationism, are religion, not science, and that it is therefore unconstitutional to include them in public school science classes. Below are excerpts from three of the most prominent cases.

Supreme Court of the United States, *Epperson v. Arkansas*, 1968

"Government in our democracy, state and national, must be neutral in matters of religious theory, doctrine, and practice. It may not be hostile to any religion or to the advocacy of non-religion, and it may not aid, foster, or promote one religion or religious theory against another or even against the militant opposite."

Supreme Court of the United States, *Edwards v. Aguillard*, 1987

"[The] primary purpose [of the Louisiana 'Creation Act,' which required the teaching of 'creation science' together with evolution in public schools] was to change the public school science curriculum to provide persuasive advantage to a particular religious doctrine that rejects the factual basis of evolution in its entirety. Thus, the Act is designed either to promote the theory of creation science that embodies a particular religious tenet or to prohibit the teaching of a scientific theory disfavored by certain religious sects. In either case, the Act violates the First Amendment."

U.S. District Court for the Middle District of Pennsylvania,
*Kitzmiller v. Dover Area School District*, 2005

"[W]e find that ID [intelligent design] is not science and cannot be adjudged a valid, accepted scientific theory, as it has failed to publish in peer-reviewed journals, engage in research and testing, and gain acceptance in the scientific community.  ID, as noted, is grounded in theology, not science. . . .  Moreover, ID's backers have sought to avoid the scientific scrutiny which we have now determined that it cannot withstand by advocating that the controversy, but not ID itself, should be taught in science class.  This tactic is at best disingenuous, and at worst a canard.  The goal of the IDM [intelligent design movement] is not to encourage critical thought, but to foment a revolution which would supplant evolutionary theory with ID."

U.S. law does not forbid the mention or study of religion as an academic subject in public schools, and creationism might be discussed in, for example, a comparative religion class. But, as civil servants, public school teachers must be neutral with respect to religion, which means that they can neither promote nor inhibit its practice. If intelligent design creationism were to be discussed in public school, then Hindu, Islamic, Native American, and other non-Christian creationist views, as well as mainstream religious views that are compatible with science, also should be discussed. Because the Constitution of the United States forbids a governmental establishment of religion, it would be inappropriate to use public funds to teach the views of just one religion or one religious subgroup to all students. Moreover, even in such a class it would be improper to teach these viewpoints as though they were scientific.

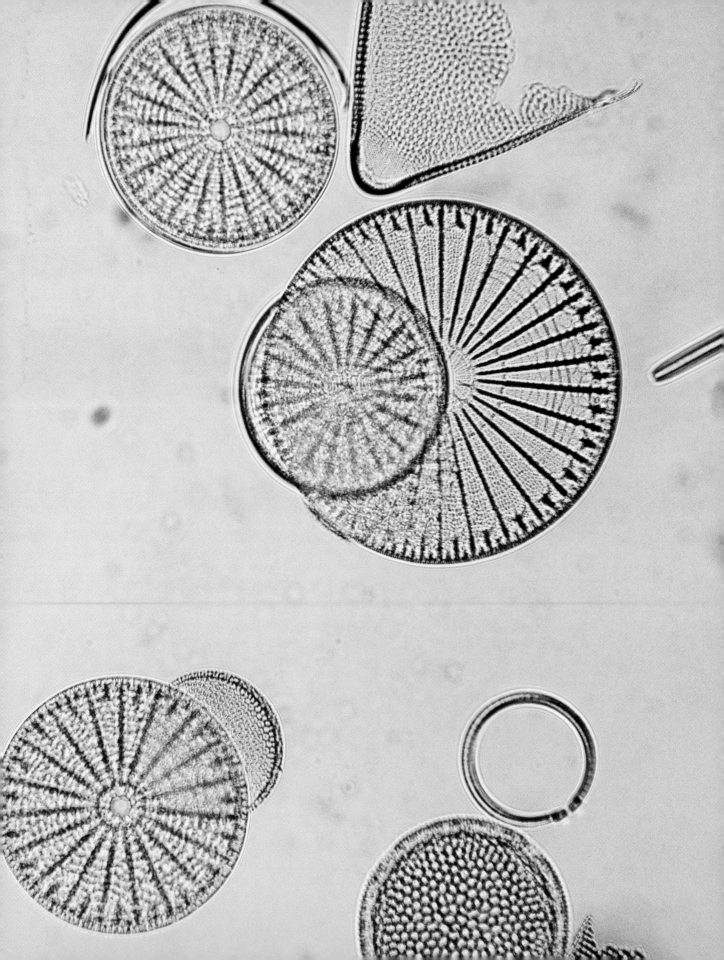

CHAPTER FOUR

# CONCLUSION

Science and science-based technologies have transformed modern life. They have led to major improvements in living standards, public welfare, health, and security. They have changed how we view the universe and how we think about ourselves in relation to the world around us.

Biological evolution is one of the most important ideas of modern science. Evolution is supported by abundant evidence from many different fields of scientific investigation. It underlies the modern biological sciences, including the biomedical sciences, and has applications in many other scientific and engineering disciplines.

As individuals and societies, we are now making decisions that will have profound consequences for future generations. How should we balance the need to preserve the Earth's plants, animals, and natural environment against other pressing concerns? Should we alter our use of fossil fuels and other natural resources to enhance the well-being of our descendants? To what extent should we use our new understanding of biology on a molecular level to alter the characteristics of living things?

None of these decisions can be made wisely without considering biological evolution. People need to understand evolution, its role within the broader scientific enterprise, and its vital implications for some of the most pressing social, cultural, and political issues of our time.

Science and technology are so pervasive in modern society that students increasingly need a sound education in the core concepts, applications, and implications of science. Because evolution has and will continue to serve as a critical foundation of the biomedical and life sciences, helping students learn about and understand the scientific evidence, mechanisms, and implications of evolution are fundamental to a high-quality science education.

Science and religion are different ways of understanding. Needlessly placing them in opposition reduces the potential of both to contribute to a better future.

# FREQUENTLY ASKED QUESTIONS

## Aren't evolution and religion opposing ideas?

Newspaper and television stories sometimes make it seem as though evolution and religion are incompatible, but that is not true. Many scientists and theologians have written about how one can accept both faith and the validity of biological evolution. Many past and current scientists who have made major contributions to our understanding of the world have been devoutly religious. At the same time, many religious people accept the reality of evolution, and many religious denominations have issued emphatic statements reflecting this acceptance. (For more information, see http://www.ncseweb. org/resources/articles/1028_statements_from_religious_org_12_19_2002.asp.)

To be sure, disagreements do exist. Some people reject any science that contains the word "evolution"; others reject all forms of religion. The range of beliefs about science and about religion is very broad. Regrettably, those who occupy the extremes of this range often have set the tone of public discussions. Evolution is science, however, and only science should be taught and learned in science classes.

The "Additional Readings" section of this publication cites a number of books and articles that explore in depth the intersection of science and faith.

## Isn't belief in evolution also a matter of faith?

*Acceptance* of evolution is not the same as a religious *belief*. Scientists' confidence about the occurrence of evolution is based on an overwhelming amount of supporting evidence gathered from many aspects of the natural world. To be accepted, scientific knowledge has to withstand the scrutiny of testing, retesting, and experimentation. Evolution is accepted within the scientific community because the concept has withstood extensive testing by many thousands of scientists for more than a century. As a 2006 "Statement on the Teaching of Evolution" from the Interacademy Panel on International Issues, a global network of national science academies, said, "*Evidence-based* facts about the origins and evolution of the Earth and of life on this planet have been established by numerous observations and independently derived experimental results from a multitude of scientific disciplines" (emphasis in original). (See http://www.interacademies.net/Object.File/Master/6/150/Evolution%20statement.pdf.)

Many religious beliefs do not rely on evidence gathered from the natural world. On the contrary, an important component of religious belief is faith, which implies acceptance of a truth regardless of the presence of empirical evidence for or against that truth. Scientists cannot accept scientific conclusions on faith alone because all such conclusions must be subject to testing against observations. Thus, scientists do not "believe" in evolution in the same way that someone believes in God.

## How can random biological changes lead to more adapted organisms?

Contrary to a widespread public impression, biological evolution is not random, even though the biological changes that provide the raw material for evolution are not directed toward predetermined, specific goals. When DNA is being copied, mistakes in the copying process generate novel DNA sequences. These new sequences act as evolutionary "experiments." Most mutations do not change traits or fitness. But some mutations give organisms traits that enhance their ability to survive and reproduce, while other mutations reduce the reproductive fitness of an organism.

The process by which organisms with advantageous variations have greater reproductive success than other organisms within a population is known as "natural selection." Over multiple generations, some populations of organisms subjected to natural selection may change in ways that make them better able to survive and reproduce in a given environment. Others may be unable to adapt to a changing environment and will become extinct.

## Aren't there many questions that still surround evolution? Don't many famous scientists reject evolution?

As with *all* active areas of science, there remain questions about evolution. There are always new questions to ask, new situations to consider, and new ways to study known phenomena. But evolution itself has been so thoroughly tested that biologists are no longer examining *whether* evolution has occurred and is continuing to occur. Similarly, biologists no longer debate many of the mechanisms responsible for evolution. As with any other field of science, scientists continue to study the *mechanisms* of how the process of evolution operates. As new technologies make possible previously unimaginable observations and allow for new kinds of experiments, scientists continue to propose and examine the strength of evidence regarding the mechanisms for evolutionary change. But the existence of such questions neither reduces nor undermines the fact that evolution has occurred and continues to occur.

Nor do such questions diminish the strength of evolutionary science. Indeed, the strength of a theory rests in part on providing scientists with the basis to explain observed phenomena and to predict what they are likely to find when exploring new phenomena and observations. In this regard, evolution has been and continues to be one of the most productive theories known to modern science.

Even scientific theories that are firmly established continue to be tested and modified by scientists as new information and new technologies become available. For example, the theory of gravity has been substantiated by many observations on Earth. But theoretical scientists, using their understanding of the physical universe, continue to test the limits of the theory of gravity in more extreme situations, such as close to a neutron star or black hole. Someday, new phenomena may be discovered that will require that the theory be expanded or revised, just as the development of the theory of general relativity in the first part of the 20th century expanded knowledge about gravity.

With evolutionary theory, many new insights will emerge as research proceeds. For example, the links between genetic changes and alterations in an organism's form and function are being intensively investigated now that the tools and technologies to do so are available.

Some who oppose the teaching of evolution sometimes use quotations from prominent scientists out of context to claim that scientists do not support evolution. However, examination of the quotations reveals that the scientists are actually disputing some aspect of *how* evolution occurs, not *whether* evolution occurred.

## What evidence is there that the universe is billions of years old?

This is an important question because evolution of the wide variety of organisms currently existing on Earth required a very long period of time. Several independent dating techniques indicate that the Earth is billions of years old. Measurements of the radioactive elements in materials from the Earth, the Moon, and meteorites provide ages for the Earth and the solar system. These measurements are consistent with each other and with the physical processes of radioactivity. Additional evidence for the ages of the solar system and the galaxy includes the record of crater formation on the planets and their moons, the ages of the oldest stars in the Milky Way, and the rate of expansion of the universe. Measurements of the radiation left over from the Big Bang also support the universe's great age.

## What's wrong with teaching critical thinking or "controversies" with regard to evolution?

Nothing is wrong with teaching critical thinking. Students need to learn how to reexamine their ideas in light of observations and accepted scientific concepts. Scientific knowledge itself is the result of the critical thinking applied by generations of scientists to questions about the natural world. Scientific knowledge must be subjected to continued reexamination and skepticism for human knowledge to continue to advance.

But critical thinking does not mean that all criticisms are equally valid. Critical thinking has to be based on rules of reason and evidence. Discussion of critical thinking or controversies does not mean giving equal weight to ideas that lack essential supporting evidence. The ideas offered by intelligent design creationists are not the products of scientific reasoning. Discussing these ideas in science classes would not be appropriate given their lack of scientific support.

Recent calls to introduce "critical analysis" into science classes disguise a broader agenda. Other attempts to introduce creationist ideas into science employ such phrases as "teach the controversy" or "present arguments for and against evolution." Many such calls are directed specifically at attacking the teaching of evolution or other topics that some people consider as controversial. In this way, they are intended to introduce creationist ideas into science classes, even though scientists have thoroughly refuted these ideas. Indeed, the application of critical thinking to the science curriculum would argue against including these ideas in science classes because they do not meet scientific standards.

There is no scientific controversy about the basic facts of evolution. In this sense the intelligent design movement's call to "teach the controversy" is unwarranted. Of course, there remain many interesting questions about evolution, such as the evolutionary origin of sex or different mechanisms of speciation, and discussion of these questions is fully warranted in science classes. However, arguments that attempt to confuse students by suggesting that there are fundamental weaknesses in the science of evolution are unwarranted based on the overwhelming evidence that supports the theory. Creationist ideas lie outside of the realm of science, and introducing them in science courses has been ruled unconstitutional by the U.S. Supreme Court and other federal courts.

## What are common ideas regarding creationism?

"Creationism" is a very broad term. In the most general sense, it refers to views that reject scientific explanations of certain features of the natural world (whether in biology, geology, or other sciences) and instead posit direct intervention (sometimes called "special creation") in these features by some transcendent being or power. Some creationists believe that the universe and Earth are only several thousand years old, a position referred to as "young Earth" creationism. Creationism also includes the view that the complex features of organisms cannot be explained by natural processes but require the intervention of a nonnatural "intelligent designer." The "Additional Readings" section following these questions contains several books that describe the various ways in which the word "creationism" is used.

## Wouldn't it be "fair" to teach creationism along with evolution?

The goal of science education is to expose students to the best possible scholarship in each field of science. The science curriculum is thus the product of centuries of scientific investigation. Ideas need to become part of the base of accepted scientific knowledge before they are appropriately taught in schools. For example, the idea of continental drift to explain the movements and shapes of the continents was studied and debated for many years without becoming part of the basic science curriculum. As data accumulated, it became clearer that the surface of the Earth is composed of a series of massive plates, which are not bounded by the continents, that continually move in relation to each other. The theory of plate tectonics (which was proposed in the mid-1960s) grew from these data and offered a more complete explanation for the movement of continents. The new theory also predicted important phenomena, such as where earthquakes and volcanoes are likely to occur. When enough evidence had accumulated for the concept of plate tectonics to be accepted by the scientific community as fact, it became part of the earth sciences curriculum.

Scientists and science educators have concluded that evolution should be taught in science classes because it is the only tested, comprehensive scientific explanation for the nature of the biological world today that is supported by overwhelming evidence and widely accepted by the scientific community. The ideas supported by creationists, in contrast, are not supported by evidence and are not accepted by the scientific community.

Different religions hold very different views and teachings about the origins and diversity of life on Earth. Because creationism is based on specific sets of religious convictions, teaching it in science classes would mean imposing a particular religious view on students and thus is unconstitutional, according to several major rulings in federal district courts and the Supreme Court of the United States.

## Does science disprove religion?

Science can neither prove nor disprove religion. Scientific advances have called some religious beliefs into question, such as the ideas that the Earth was created very recently, that the Sun goes around the Earth, and that mental illness is due to possession by spirits or demons. But many religious beliefs involve entities or ideas that currently are not within the domain of science. Thus, it would be false to assume that *all* religious beliefs can be challenged by scientific findings.

As science continues to advance, it will produce more complete and more accurate explanations for natural phenomena, including a deeper understanding of biological evolution. Both science and religion are weakened by claims that something not yet explained scientifically must be attributed to a supernatural deity. Theologians have pointed out that as scientific knowledge about phenomena that had been previously attributed to supernatural causes increases, a "god of the gaps" approach can undermine faith. Furthermore, it confuses the roles of science and religion by attributing explanations to one that belong in the domain of the other.

Many scientists have written eloquently about how their scientific studies have increased their awe and understanding of a creator (see the "Additional Readings" section). The study of science need not lessen or compromise faith.

# ADDITIONAL READINGS

## Websites

The National Academy of Sciences maintains a website containing publications and other resources from the academies that focus on evolution and evolution education. The website also contains links to other useful websites about evolution and the nature of science that are maintained by other scientific organizations. For more information see http://nationalacademies.org/evolution.

## Articles on Research Described in This Publication

Alemseged, Z., F. Spoor, W.H. Kimbel, R. Bobe, D. Geraads, D. Reed, and J.G. Wynn. 2006. A juvenile early hominid skeleton from Dikika, Ethiopia. *Nature* 443 (7109): 296–301.

Allwood, A.C., M.R. Walter, B.S. Kamber, C.P. Marshall, and I.W. Burch. 2006. Stromatolite reef from the Early Archaean era of Australia. *Nature* 441 (7094): 714–18.

Banzhaf, W., G. Beslon, S. Christensen, J.A. Foster, F. Kepes, V. Lefort, J.F. Miller, M. Radman, and J.J. Ramsden. 2006. From artificial evolution to computational evolution: a research agenda. *Nature Reviews Genetics* 7 (9): 729–35.

Bull, J.J., and H.A. Wichman. 2001. Applied evolution. *Annual Review of Ecology and Systematics* 32: 183–217.

Carson, H.L. 1997. Sexual selection: A driver of genetic change in Hawaiian *Drosophila*. *Journal of Heredity* 88 (5): 343–52.

Craddock, E.M. 2000. Speciation processes in the adaptive radiation of Hawaiian plants and animals. *Evolutionary Biology* 31: 1–43.

Daeschler, E.B., N.H. Shubin, and F.A. Jenkins Jr. 2006. A Devonian tetrapod-like fish and the evolution of the tetrapod body plan. *Nature* 440 (7085): 757–63.

Kent, W.J., R. Baertsch, A. Hinrichs, W. Miller, and D. Haussler. 2003. Evolution's cauldron: duplication, deletion, and rearrangement in the mouse and human genomes. *Proceedings of the National Academy of Sciences USA* 100 (20): 11484–89.

Ksiazek, T.G., D. Erdman, C.S. Goldsmith, S.R. Zaki, T. Peret, S. Emery, S. Tong, C. Urbani, J.A. Comer, W. Lim, P.E. Rollin, S.F. Dowell, A.E. Ling, C.D. Humphrey, W.J. Shieh, J. Guarner, C.D. Paddock, P. Rota, B. Fields, J. DeRisi, J.Y. Yang, N. Cox, J.M. Hughes, J.W. LeDuc, W.J. Bellini, L.J. Anderson— SARS Working Group. 2003. A novel coronavirus associated with severe acute respiratory syndrome. *New England Journal of Medicine* 348 (20): 1953–66.

Miller, J.D., E.D. Scott, and S. Okamoto. 2006. Public acceptance of evolution. *Science* 313 (5788): 765–66.

Reznick, D.N., F.H. Shaw, F.H. Rodd, and R.G. Shaw. 1997. Evaluation of the rate of evolution in natural populations of guppies (*Poecilia reticulata*). *Science* 275 (5308): 1934–37.

Salamini, F., H. Ozkan, A. Brandolini, R. Schafer-Pregl, and W. Martin. 2002. Genetics and geography of wild cereal domestication in the Near East. *Nature Reviews Genetics* 3 (6): 429–41.

Shubin, N.H., E.B. Daeschler, and F.A. Jenkins Jr. 2006. The pectoral fin of *Tiktaalik roseae* and the origin of the tetrapod limb. *Nature* 440 (7085): 764–71.

Thewissen, J.G., M.J. Cohn, L.S. Stevens, S. Bajpai, J. Heyning, and W.E. Horton Jr. 2006. Developmental basis for hind-limb loss in dolphins and origin of the cetacean bodyplan. *Proceedings of the National Academy of Sciences USA* 103 (22): 8414–18.

Thewissen, J.G., E.M. Williams, L.J. Roe, and S.T. Hussain. 2001. Skeletons of terrestrial cetaceans and the relationship of whales to artiodactyls. *Nature* 413 (6853): 277–81.

You, H.L., M.C. Lamanna, J.D. Harris, L.M. Chiappe, J. O'Connor, S.A. Ji , J.C. Lu, C.X. Yuan, D.Q. Li, X. Zhang, K.J. Lacovara, P. Dodson, and Q. Ji. 2006. A nearly modern amphibious bird from the Early Cretaceous of northwestern China. *Science* 312 (5780): 1640–43.

## Books on Evolution, the Nature of Science, and Science Education

The following list of references represents a sampling of the vast literature available on evolution, science, and science education. Please visit our World Wide Web address, http://nationalacademies.org/evolution, for more extensive resource listings for these subjects. The National Academy of Sciences does not endorse all the views or perspectives expressed by the authors of the following books.

### Books on Evolution

Ayala, Francisco J. 2007. *Darwin's Gift to Science and Religion*. Washington, DC: Joseph Henry Press.

An evolutionary biologist with a background in theology explains the science of evolution and its compatibility with faith.

Carroll, Sean B. 2006. *The Making of the Fittest: DNA and the Ultimate Forensic Record of Evolution*. New York: Norton.

Links changes in DNA over time to the evolution of organisms and explores the new science of evolutionary-development biology, or "evo-devo."

Dawkins, Richard. 1996. *Climbing Mount Improbable*. New York: Norton.

An authoritative and elegant account of the evolutionary origins of the "design" of organisms.

Dennett, Daniel C. 1995. *Darwin's Dangerous Idea: Evolution and the Meanings of Life*. New York: Simon and Schuster.

An exploration of Darwin's conceptual advances and of how those advances have influenced other areas of thought.

Fortey, Richard. 1998. *Life: A Natural History of the First Four Billion Years of Life on Earth*. New York: Knopf.

A lively account of the history of life on Earth.

Gould, Stephen J. 1992. *The Panda's Thumb*. New York: W.W. Norton.

Gould's columns from *Natural History* magazine have been collected into a series of books, including *Ever Since Darwin, Hen's Teeth and Horses' Toes, Eight Little Piggies, The Flamingo's Smile,* and *Bully for Brontosaurus.* All are good popular introductions to the basic ideas behind evolution, and extremely readable.

Hazen, Robert M. 2006. *Genesis: The Scientific Quest for Life's Origins*. Washington, DC: Joseph Henry Press.

An engaging introduction to current ideas about the origin of life on Earth, featuring vivid descriptions of the experiments Hazen and others are doing to test possible mechanisms.

Horner, John R., and Edwin Dobb. 1997. *Dinosaur Lives: Unearthing an Evolutionary Saga*. New York: Harper Collins.

What it's like to uncover fossilized bones, eggs, and more, plus Horner's views on dinosaurs and evolution.

Kirschner, Marc W., and John C. Gerhart. 2005. *The Plausibility of Life: Resolving Darwin's Dilemma*. New Haven, CT: Yale University Press.

Explains how small changes in the DNA of an organism can produce new biological structures and systems.

Mayr, Ernst. 2001. *What Evolution Is*. New York: Basic Books.

An authoritative and comprehensive overview of evolutionary theory.

Mindell, David P. 2006. *The Evolving World: Evolution in Everyday Life*. Cambridge, MA: Harvard University Press.

Describes the many practical applications of evolutionary knowledge in medicine, agriculture, conservation biology, and other fields.

National Academy of Sciences. 1998. *Teaching About Evolution and the Nature of Science*. Washington, DC: National Academy Press.

A guide for educators, policy makers, parents, and others that offers guidance on teaching about evolution and the nature of science.

Weiner, Jonathan. 1994. *The Beak of the Finch: A Story of Evolution in Our Time*. New York: Knopf.

Discussion of basic evolutionary principles and how they are illustrated by ongoing evolution of the finches on the Galápagos Islands.

Zimmer, Carl. 2002. *Evolution: The Triumph of an Idea*. New York: Harper.

A broad overview of evolution — and companion to a PBS series of the same name — that examines the influence and scope of Darwin's ideas.

## Books on the Evolution of Humans

Cela-Conde, Camilo J., and Francisco J. Ayala. 2007. *Human Evolution: Trails from the Past.* New York: Oxford University Press.

A comprehensive overview of the evolution of humans that draws from fields ranging from genomics and paleoanthropology to ethics and religion.

Diamond, Jared. 1993, reissued in 2006. *The Third Chimpanzee: The Evolution and Future of the Human Animal.* New York: Harper Perennial.

Discusses the similarities and differences between humans and chimpanzees.

Howells, William W. 1997. *Getting Here: The Story of Human Evolution.* Washington, DC: Compass Press.

A readable survey of human evolution by one of the fathers of physical anthropology.

Stringer, Chris, and Peter Andrews. 2005. *The Complete World of Human Evolution.* New York: Thames and Hudson.

A thorough, well-illustrated, and up-to-date guide to the evolution of human beings from our nonhuman ancestors.

Tattersall, Ian. 1998. *Becoming Human: Evolution and Human Uniqueness.* New York: Harcourt Brace.

A description of the current state of understanding about the differences between Neanderthals and modern humans.

Zimmer, Carl. 2005. *Smithsonian Intimate Guide to Human Origins.* Washington, DC: Smithsonian Books.

A succinct guide to the complex story of human evolution.

## Books on Evolution for Children and Young Adults

Jenkins, Steve. 2002. *Life on Earth: The Story of Evolution.* Boston: Houghton Mifflin.

A remarkably broad and detailed introduction to evolutionary theory. Grades 2–6.

Lauber, Patricia. 1994. *How Dinosaurs Came to Be.* New York: Simon and Schuster.

A description of the dinosaurs and their ancestors. Grades 4–7.

Lawson, Kristan. 2003. *Darwin and Evolution for Kids: His Life and Ideas with 21 Activities.* Chicago: Chicago Review.

A life of Darwin combined with activities such as making a taxonomy and investigating geological strata. Grades 5–9.

Matsen, Bradford. 1994. *Planet Ocean: A Story of Life, the Sea, and Dancing to the Fossil Record.* Berkeley, CA: Ten Speed Press.

Whimsically illustrated tour of history for older kids and adults. Grades 6–10.

McNulty, Faith. 1999. *How Whales Walked into the Sea.* New York: Scholastic.

This wonderfully illustrated book describes the evolution of whales from land mammals. Grades K–5.

Peters, Lisa W. 2003. *Our Family Tree: An Evolution Story.* New York: Harcourt.

A beautifully illustrated picture book that emphasizes the relatedness of all living things. Grades K–5.

Troll, Ray, and Bradford Matsen. 1996. *Raptors, Fossils, Fins & Fangs: A Prehistoric Creature Feature.* Berkeley, CA: Tricycle Press.

A light-hearted trip through time ("Good Gracious—Cretaceous!"). Grades 3–6.

## Books on the Origin of the Universe and the Earth

Astronomy Education Board. 2004. *An Ancient Universe: How Astronomers Know the Vast Scale of Cosmic Time.* Washington, DC: American Astronomical Society and Astronomical Society of the Pacific.

A guide for teachers, students, and the public to the methods astronomers have used to date the cosmos.

Dalrymple, G. Brent. 2004. *Ancient Earth, Ancient Skies: The Age of Earth and Its Cosmic Surroundings.* Palo Alto, CA: Stanford University Press.

A comprehensive discussion of the evidence for the ages of the Earth, Moon, meteorites, solar system, galaxy, and universe.

Longair, Malcolm S. 2006. *The Cosmic Century: A History of Astrophysics and Cosmology.* New York: Cambridge.

A review of the historical development of astrophysics and cosmology, with an emphasis on the theoretical concepts that tie these fields to other areas of science.

Tyson, Neil D. 2007. *Death by Black Hole: And Other Cosmic Quandaries.* New York: W. W. Norton.

A collection of essays from Tyson's monthly "Universe" column in *Natural History* magazine on how science works and how we have come to understand our place in the universe.

Tyson, Neil D., and Donald Goldsmith. 2004. *Origins: Fourteen Billion Years of Cosmic Evolution.* New York: W. W. Norton.

The companion book to the NOVA series "Origins," conveys the latest understanding of the origin of the universe, galaxies, stars, planets, and life.

Weinberg, Steven. 1993. *The First Three Minutes: A Modern View of the Origin of the Universe.* New York: Basic Books.

An explanation of what happened during the Big Bang.


**Books on Genomics and Evolution**

DeSalle, Rob, and Michael Yudell. 2004. *Welcome to the Genome: A User's Guide to the Genetic Past, Present, and Future.* New York: Wiley.

Discusses the science, the applications, and the potential of human genetics.

Ridley, Matt. 1999. *Genome: The Autobiography of a Species in 23 Chapters.* New York: HarperCollins.

A chromosome-by-chromosome investigation of how genetics research could change human life.

Watson, James D., and Andrew Berry. 2003. *DNA: The Secret of Life.* New York: Knopf.

A history of genetics written in part by the scientist who uncovered the structure of DNA.


**Books on the Evolution and Creationism Controversy**

Ayala, Francisco J. 2006. *Darwin and Intelligent Design.* Minneapolis, MN: Fortress Press.

A comparison of evolutionary theory with the ideas proposed by the backers of "intelligent design creationism."

Baker, Catherine, and James B. Miller. 2006. *The Evolution Dialogues: Science, Christianity, and the Quest for Understanding.* Washington, DC: American Association for the Advancement of Science.

Alternating chapters present the findings of science and the Christian response to those findings in a publication generated as part of the dialogue on science, ethics, and religion sponsored by the AAAS.

Collins, Francis. 2006. *The Language of God: A Scientist Presents Evidence for Belief.* New York: Free Press.

The director of the Human Genome Project describes his religious beliefs in the context of his scientific research.

Forrest, Barbara, and Paul R. Gross. 2004. *Creationism's Trojan Horse: The Wedge of Intelligent Design.* New York: Oxford University Press.

A close analysis of the positions and tactics taken by the intelligent design branch of creationism.

Humes, Edward. 2007. *Monkey Girl: Evolution, Education, Religion, and the Battle for America's Soul.* New York: HarperCollins.

An eyewitness account of the *Kitzmiller vs. Dover Area School District* trial.

Kitcher, Philip. 2006. *Living with Darwin: Evolution, Design, and the Future of Faith.* New York: Oxford University Press.

A philosopher of science compares different versions of creationism to evolution while examining the broader differences between religious and scientific perspectives.

Matsumura, Molleen. 1995. *Voices for Evolution.* Berkeley, CA: National Center for Science Education. Continually updated at http://www.ncseweb.org/article.asp?category=2.

A collection of statements supporting the teaching of evolution from many different types of organizations: scientific, civil liberties, religious, and educational.

Miller, Kenneth R. 1999. *Finding Darwin's God: A Scientist's Search for Common Ground Between God and Evolution.* New York: HarperCollins.

A biologist seeks to reconcile evolutionary theory with a belief in God.

Moore, John A. 2002. *From Genesis to Genetics: The Case of Evolution and Creationism.* Berkeley, CA: University of California Press.

An argument for the educational importance of teaching evolution.

Nelkin, Dorothy. 2000. *The Creation Controversy: Science or Scripture in Schools.* Lincoln, NE: iUniverse, Inc.

A sociologist of science examines the controversies in Kansas about teaching evolution and questions about the public's trust in science.

Pennock, Robert T. 1999. *Tower of Babel: The Evidence Against the New Creationism.* Cambridge, MA: MIT Press.

A philosopher of science analyzes "intelligent design" and "theistic science" creationism.

Pennock, Robert T., ed. 2001. *Intelligent Design Creationism and Its Critics: Philosophical, Theological, and Scientific Perspectives.* Cambridge, MA: MIT Press.

A collection of papers by creationists and their critics, with a particular focus on "intelligent design creationism."

Pigliucci, Massimo. 2002. *Denying Evolution: Creationism, Scientism, and the Nature of Science.* Sunderland, MA: Sinauer Associates.

Examines the history of the evolution/creationism "debate" and provides detailed information about what the author sees as fallacies by both creationists and scientists.

Ruse, Michael. 2005. *The Evolution-Creation Struggle.* Cambridge, MA: Harvard University Press.

A history of the reaction to Darwin's ideas that offers constructive suggestions for advocates on both sides of the debate.

Scott, Eugenie. 2005. *Evolution vs. Creationism: An Introduction.* Berkeley, CA: University of California Press.

Written by the executive director of the National Center for Science Education, this survey of the issues surrounding debates over evolution and creationism includes useful lists of primary documents and resources.

Scott, Eugenie, and Glenn Branch, eds. 2006. *Not in Our Classrooms: Why Intelligent Design Is Wrong for Our Schools.* Boston, MA: Beacon Press.

A collection of essays that examines the history of "intelligent design creationism" and the legal controversies surrounding its introduction into public school classrooms.

# COMMITTEE MEMBER BIOGRAPHIES

**Bruce Alberts** (NAS) is professor of biochemistry and biophysics at the University of California, San Francisco. His research has focused on the mechanisms of two different reactions that are fundamental to the life of the cell. He is noted particularly for his extensive study of the protein complexes that allow chromosomes to be replicated, as required for a living cell to divide.

Alberts is one of the original authors of *The Molecular Biology of the Cell*, considered the field's leading advanced textbook and used widely in U.S. colleges and universities. His most recent text, *Essential Cell Biology*, is intended to present this subject matter to a wider audience.

He was president of the National Academy of Sciences and chair of the National Research Council from 1993 to 2005. He continues to serve as an ex officio member of the National Academies Teacher Advisory Council, which he initiated. Committed to improving science education, he helped initiate and develop City Science, a program that links UCSF to the improvement of science teaching in San Francisco elementary schools.

**Francisco J. Ayala** (Committee Chair, NAS) is university professor and Donald Bren professor of biological sciences and professor of philosophy at the University of California, Irvine. His research focuses on population and evolutionary genetics. The study of biological evolution is his main interest, particularly the genetics of the evolutionary process, molecular evolution, the process of speciation, genetic variation in populations, studies of population growth and dynamics, and ecological competition. He also writes about the interface between religion and science, and on philosophical issues concerning epistemology, ethics, and the philosophy of biology. His books include *Human Evolution: Trails from the Past*, *Darwin's Gift to Science and Religion*, *Darwin and Intelligent Design*, *Population and Evolutionary Genetics: A Primer*, *Evolving: The Theory and Processes of Organic Evolution*, and *Studies in the Philosophy of Biology*. He testified in the Arkansas trial on the teaching of evolution in 1981.

He has been president and chairman of the board of the American Association for the Advancement of Science and president of Sigma Xi, the scientific research society of the United States. He has received awards from many organizations worldwide, as well as honorary degrees from universities in Europe, Asia, and the United States. In 2002, President George W. Bush awarded him the National Medal of Science.

**May R. Berenbaum** (NAS) is the Swanlund professor and head of the Department of Entomology at the University of Illinois at Urbana-Champaign. She has made major contributions to understanding the role of chemistry in interactions between plants and herbivorous insects and identifying key plant toxins and determining their modes of action against insects. Her investigations have examined proximate physiological mechanisms and their evolutionary consequences for both plants and insects. Her research interests include chemical ecology, insect-plant interactions, the evolutionary biology of moths and butterflies (*Lepidoptera*), photobiology, and environmentally sustainable pest management.

She has received awards from the National Science Foundation, the Ecological Society of America, the Weizmann Institute, and the International Society of Chemical Ecology. She is an elected fellow of the Entomological Society of America, the American Academy of Arts and Sciences, and the American Philosophical Society.

She is a member of the editorial board of the *Proceedings of the National Academy of Sciences* and a recent member of the Council of the National Academy of Sciences. As a result of her interest in promoting science literacy, she has authored many newspaper and magazine articles and four books on science topics for general readers.

**Betty Carvellas** is a recently retired teacher and science department cochair at Essex High School in Essex Junction, Vermont. Her professional service included work at the local, state, and national levels. She served as cochair of the education committee and was a member of the executive board of the Council of Scientific Society Presidents and is a past president of the National Association of Biology Teachers.

She received the Sigma Xi Outstanding Vermont Science Teacher Award (1981) and the Presidential Award for Excellence in Science and Mathematics Teaching (1984), and in 2000 she was named honorary member of the National Association of Biology Teachers. In 2001 she was selected for a National Science Foundation program, Teachers Experiencing Antarctica and the Arctic, and she has spent four summers working with scientists in the Bering Sea and the Arctic Ocean. She was a charter member and chair of the Vermont Standards Board for Professional

Educators and served on the board of directors of the Biological Sciences Curriculum Study.

Her interests include interdisciplinary teaching, connecting school science to the real world, traveling with students on international field studies, and bringing inquiry into the science classroom. Carvellas was a charter member of the Teacher Advisory Council of the National Academies, and she served as chair of the ad hoc committee that organized its 2004 workshop on linking mandatory professional development to high-quality teaching and learning.

**Michael T. Clegg** (NAS) is Donald Bren professor of biological sciences at the University of California, Irvine. He is an authority on the evolution of complex genetic systems and is recognized internationally for his contributions to understanding the genetic and ecological basis for adaptive evolutionary changes in populations and at higher taxonomic levels. He is interested in the population genetics of plants, plant molecular evolution, statistical estimation of genetic parameters, plant phylogeny, plant genetic transmission and molecular genetics, and genetic conservation in agriculture.

Clegg is an ex officio member of 29 National Academy of Sciences committees, as well as chair of the International Advisory Board and a member of the International Programs Committee. He is currently serving as foreign secretary of the National Academy of Sciences. He chaired the delegation to the 28th General Assembly of the International Council for Science in Shanghai and Suzhou, China, in 2005.

**G. Brent Dalrymple** (NAS) is professor and dean emeritus of oceanic and atmospheric sciences at Oregon State University. He is a geochronologist who helped lay the basis for ocean-floor spreading theory, the hotspot theory of mid-ocean volcanism, the use of mantle plumes as the absolute frame for plate motion through geologic history, fine-structure stratigraphy of the lunar regolith, and lunar cratering history. His primary research interests involve the development and improvement of isotopic dating techniques, in particular the K-Ar and $^{40}Ar/^{39}Ar$ methods, and their application to a broad range of geological and geophysical problems.

Dalrymple is the author of *The Age of the Earth* as well as a shorter version titled *Ancient Earth, Ancient Skies.* His recent research involves a series of experiments to determine the history of bombardment of the Moon by large impactors and of the resulting lunar basin formation. He testified in the landmark federal cases on evolution education, *McLean v. Arkansas* and *Aguillard v. Treen.*

He is a fellow of the American Geophysical Union, serving as president and a member of the board of directors, and the American Academy of Arts and Sciences. He received the 2001 Public Service Award from the Geological Society of America and the 2003 National Medal of Science.

**Robert M. Hazen** is a research scientist at the Carnegie Institution of Washington's Geophysical Laboratory and the Clarence Robinson professor of earth science at George Mason University. His recent research focuses on the role of minerals in the origin of life, including such processes as mineral-catalyzed organic synthesis and the selective adsorption of organic molecules on mineral surfaces. He is the author of *Genesis: The Scientific Quest for Life's Origins, The New Alchemists, Why Aren't Black Holes Black?, The Diamond Makers,* and more than 260 scientific papers.

Hazen is active in presenting science to a general audience. At George Mason University he has developed courses and companion texts on scientific literacy. His books with coauthor James Trefil include *Science Matters: Achieving Scientific Literacy* and *The Sciences: An Integrated Approach.* He also served on the team of writers for the NRC's National Science Education Standards and the National Academy's *Teaching About Evolution and the Nature of Science.*

He serves on the Committee on Public Understanding of Science and Technology of the American Association for the Advancement of Science and on advisory boards for NOVA (WGBH, Boston), Earth & Sky (PBS), the Encyclopedia Americana, and the Carnegie Council. He appears frequently on radio and television programs on science, and he recorded *The Joy of Science,* a 60-lecture video course produced by The Teaching Company.

He was recently elected president of the Mineralogical Society of America. A fellow of the American Association for the Advancement of Science, he has received awards from the Mineralogical Society of America, the American Chemical Society, the American Society of Composers, Authors, and Publishers, the Educational Press Association, and the American Crystallographic Association.

**Toby M. Horn** is codirector of the Carnegie Academy for Science Education at the Carnegie Institution of Washington, D.C. In this capacity she works directly with teachers in the District of Columbia public schools, both in workshops and in their classrooms, to help them improve instruction in science, mathematics, and technology. She also works with the D.C. public school system to assist teachers in obtaining the necessary supplies for teaching science and biotechnology. Horn is an instructor in the academy's First Light Saturday science program for middle school students in D.C. public and charter schools.

Prior to joining Carnegie, Horn taught at the Thomas Jefferson High School for Science and Technology in Fairfax County, Va., and established one of the first precollege biotechnology programs there. She also served for two years as outreach coordinator for the Fralin Biotechnology Center at Virginia Polytechnic Institute and State University.

Horn was the 2006 president of the National Association of Biology Teachers. As a staff fellow at the National Cancer Institute, she studied DNA sequences thought to be associated with breast cancer.

**Nancy A. Moran** (NAS) is Regents' professor of ecology and evolutionary biology at the University of Arizona. She is active in interdisciplinary graduate training in evolutionary genomics and has taught evolutionary biology and genomics at the undergraduate, graduate, and high school levels. Her research focuses on the role of symbiotic interactions in ecology and evolution and involves fundamental evolutionary forces, such as mutation, gene transfer, natural selection, and ecological diversification. Using approaches from molecular evolution, systematics, genomics, and population genetics, she works extensively with both bacteria and insects and their ecological interactions. Her work has shown that many groups of insects have coevolved with bacterial symbionts for millions of years, that these symbionts supply nutrients to their hosts, allowing diversification into new ecological niches, and that the symbionts have undergone extensive genome reduction through loss of most ancestral genes. Most of her work is on groups of insects, such as aphids, that are major agricultural pests.

Moran has served as president of the Society for the Study of Evolution and as vice president of the American Society of Naturalists. She is a fellow of the American Academy of Microbiology and a recipient of a MacArthur Foundation fellowship.

**Gilbert S. Omenn** (IOM) is professor of internal medicine, human genetics, and public health and director of the Center for Computational Medicine and Biology at the University of Michigan. He is principal investigator of the Michigan Proteomics Alliance for Cancer Research and leader of the international Human Proteome Organization's Human Plasma Proteome Project.

His research interests include cancer proteomics, chemoprevention of cancers, public health genetics, science-based risk analysis, and health policy. He was principal investigator of the beta-Carotene and Retinol Efficacy Trial (CARET) of preventive agents against lung cancer and heart disease, director of the Center for Health Promotion in Older Adults, and creator of a university-wide initiative on Public Health Genetics in Ethical, Legal, and Policy Context while at the University of Washington and Fred Hutchinson Cancer Research Center. He is a longtime director of Amgen Inc. and of Rohm & Haas Company. He was president of the American Association for the Advancement of Science in 2005–2006.

He is a member of the American Academy of Arts and Sciences, the Association of American Physicians, and the American College of Physicians. He chaired the presidential/congressional Commission on Risk Assessment and Risk Management, served on the National Commission on the Environment, and chaired the National Academies' Committee on Science, Engineering, and Public Policy.

**Robert T. Pennock** is professor of history and philosophy of science at Michigan State University, where he is on the faculty of the Lyman Briggs College of Science, the Philosophy Department, and the Department of Computer Science, as well as the Center for Ethics and Humanities in the Life Sciences and the Ecology, Evolutionary Biology, and Behavior graduate program. His research interests include the philosophy of biology and the relationship of epistemic and ethical values in science.

Pennock is the author of *Tower of Babel: The Evidence Against the New Creationism* and *Intelligent Design Creationism and Its Critics: Philosophical, Theological, and Scientific Perspectives*. He testified in the case on the teaching of intelligent design creationism, *Kitzmiller v. Dover Area School District*.

Pennock has received fellowships from the Mellon Foundation, the National Endowment for the Humanities, and the National Science Foundation. He is a fellow of the American Association for the Advancement of Science and serves on its Committee on the Public Understanding of Science, as well as the American Philosophical Association's Committee on Public Philosophy. He is chair of the education committee of the Society for the Study of Evolution and is currently working on a book examining how Darwinian evolution, as an abstract theoretical model, can be applied practically beyond biology.

**Peter H. Raven** (NAS) is the Engelmann professor of botany at Washington University and director of the Missouri Botanical Garden in St. Louis. He is a conservationist who has transformed the Missouri Botanical Garden into one of the world's leading plant conservation centers. His primary research interests are the systematics, evolution, and biogeography of the plant family *Onagraceae*, which includes 16 genera and some 650 species. This family of plants has provided powerful models for understanding patterns and processes in plant evolution in general. Another particular interest is plant biogeography — the evolutionary history of entire biota and the individual taxa found in certain regions — and the ways in which these organisms have been influenced by continental movements. He has focused much of his attention on what he considers the menace of a "sixth extinction" — a potential mass extinction of living organisms that would be brought about by the mushrooming human population and by human carelessness and commerce.

Raven's service to national and international organizations has included president of the American Association for the Advancement of Science, member of the Pontifical Academy of Science, home secretary of the U.S. National Academy of Sciences, member of the President's Committee of Advisors on Science and Technology, and chairman of the National Geographic Society's Committee for Research and Exploration. He has received Guggenheim and MacArthur Foundation fellowships. *Time* magazine, in its 1999 Earth Day issue, declared that Raven is one of its "Heroes of the Planet" for what he is doing "to preserve and protect the environment."

**Barbara A. Schaal** (NAS) is the Spencer T. Olin professor of biology at Washington University, St. Louis. Her investigations have focused on the genetic heterogeneity of plant species, including those native to the United States, tropical crops and their wild relatives, and the family of plants called *Arabidopsis*. She uses a variety of molecular markers in several plant species to study fundamental evolutionary processes, such as gene migration, molecular evolution, and natural selection. Her application of DNA analysis to plant evolution at the population level has revealed unexpectedly high levels of diversity, has led to the development of DNA fingerprinting in plants, and has helped explain the reasons for this level of diversity. She has been involved with work that has identified the wild progenitor of cassava and the probable geographical location of its domestication in the Amazon region of Brazil. She has also examined the evolutionary origins of invasive plants that encroach on wetlands in the western United States. Her recent work has examined gene flow between genetically modified rice and wild relatives of rice.

Schaal currently serves as the vice president of the National Academy of Sciences. She has also been president of the Society for the Study of Evolution and the Botanical Society of America.

**Neil deGrasse Tyson** is the Frederick P. Rose director of the Hayden Planetarium at the American Museum of Natural History. His research interests include star formation, exploding stars, dwarf galaxies, and the structure of the Milky Way.

Tyson has served on presidential commissions that studied the future of the U.S. aerospace industry (2001) and the implementation of the U.S. space exploration policy (2004). A winner of the Public Service Medal of the National Aeronautics and Space Administration, the highest award given to a non–civil servant, Tyson currently serves on NASA's advisory council.

In addition to his professional publications, Tyson also writes for the public. He is an essayist for *Natural History* magazine and the author of *The Sky Is Not the Limit: Adventures of an Urban Astrophysicist* and *Origins: Fourteen Billion Years of Cosmic Evolution*, cowritten with Donald Goldsmith. He serves as the host and executive editor for the PBS-NOVA program "NOVA Science Now," in which each episode profiles the frontier of scientific discovery drawn from such fields as chemistry, biology, geology, physics, robotics, and astrophysics.

Tyson is the recipient of eight honorary doctoral degrees and currently serves as president of the Planetary Society. His contributions to public appreciation of the cosmos have recently been recognized by the International Astronomical Union in their official naming of the asteroid "13123 Tyson."

**Holly A. Wichman** is professor of biological sciences at the University of Idaho and cofounder of the interdisciplinary Initiative for Bioinformatics and Evolutionary Studies. She teaches courses in genetics, experimental biology, and professional development for graduate students. Her research focuses on genome organization in mammals and on experimental evolution using viruses as a model system. Her work on mammalian retrotransposons is carried out in a strong phylogenetic framework; she has examined retrotransposon evolution in monotremes, marsupials, and all 18 orders of placental mammals. This work focuses primarily on events that occurred tens of millions of years ago. However, short-term evolution of organisms with generation times that are short relative to that of humans can be observed in real time, both in the laboratory and in natural environments. Wichman uses the bacteriophage X174 and its relatives to study the molecular details of adaptive evolution. She studies the patterns and predictability of adaptation to novel environments such as host switching.

Wichman is also interested in applications of evolutionary biology to practical problems in industry, agriculture, and medicine. In 2001, she coauthored a comprehensive review article on applied evolution to offer examples for those who teach at the high school and undergraduate levels; it remains one of the most downloaded articles in the *Annual Review of Ecology and Systematics* series. This year she co-organized the National Institutes of Health's Workshop on Evolution of Infections Diseases and participated in the National Science Foundation's Workshop on Frontiers in Evolutionary Biology.

# STAFF AND CONSULTANT BIOGRAPHIES

**Jay B. Labov** serves as a senior advisor for education and communications for the National Academy of Sciences and the National Research Council. He also served for three years as deputy director of the National Research Council's Center for Education and was the study director and responsible staff officer for the NRC reports *Enhancing Professional Development for Teachers: Potential Uses of Information Technology, Report of a Workshop (2007); Evaluating and Improving Undergraduate Teaching in Science, Mathematics, Engineering, and Technology (2003); Learning and Understanding: Improving Advanced Study of Mathematics and Science in U.S. High Schools (2002); Educating Teachers of Science, Mathematics, and Technology: New Practices for the New Millennium (2000); Transforming Undergraduate Education in Science, Mathematics, Engineering, and Technology (1999); Serving the Needs of Pre-College Science and Mathematics Education: Impact of a Digital National Library on Teacher Education and Practice (1999);* and *Developing a Digital National Library for Undergraduate Science, Mathematics, Engineering, and Technology Education (1998).*

He also currently oversees the National Academies' activities to improve the teaching of evolution in public schools and a recently expanded effort to work more closely with disciplinary and professional societies on education issues. He has worked with many national organizations and professional societies to improve science education for both precollege and undergraduate students. He was elected as a Fellow in Education of the American Association for the Advancement of Science in 2005.

**Edward Maibach** is professor and director of the Center of Excellence in Climate Change Communication Research at George Mason University. Dr. Maibach is a highly experienced public health advocate and social change professional and a leading academic in the field of communication research. His work over the past 25 years has helped to define the fields of public health communication and social marketing, and his book *Designing Health Messages: Approaches from Communication Theory and Public Health Practice* is widely used by academics and practitioners alike. He earned his PhD in communication research from

Stanford University in 1990. He has had the pleasure of serving as Worldwide Director of Social Marketing for Porter Novelli, as an associate director of the National Cancer Institute, and in various previous academic positions.

**Steve Olson** is the author of *Mapping Human History: Genes, Race, and Our Common Origins*, a finalist for the 2002 nonfiction National Book Award and winner of the Science-in-Society Award from the National Association of Science Writers. His recent book, *Count Down: Six Kids Vie for Glory at the World's Toughest Math Competition*, was named a best science book of 2004 by *Discover* magazine. He has written several other books, including *Evolution in Hawaii* and *On Being a Scientist*. He has been a consultant writer for the National Academy of Sciences and the National Research Council, the Howard Hughes Medical Institute, the National Institutes of Health, the Institute for Genomic Research, and many other organizations.

**Barbara Kline Pope** is executive director for communications and the National Academies Press. She is responsible for an innovative and dynamic publishing operation of both scholarly and trade books that have been available on the Web free to read since 1995. Branding, marketing and audience research, derivative products, partnerships and distribution systems, and the public Web presence for the National Academies occupy her time in the communications aspects of her work. Recent research articles she has authored focus on the discipline of consumer behavior and include specific projects on business models for the digital publishing arena and the use of information sources by organizational buyers. She has been guest lecturer for marketing and technology courses at the University of Maryland and an adjunct faculty member at the University of Virginia's continuing education program. She is on the board of directors of Hands On Science Outreach, a nonprofit organization that provides high-quality after-school science programs for children.

# INDEX

## A

Adaptative radiation, 8–9, 25–28, 29
Age
    of Earth, 17, 19, 20–21, 38, 51
    of universe, 19–20, 38, 51
Amniotic eggs, 8–9
Amphibians, 23
Anthropology, 17
*Archaeopteryx*, 23
Arkansas laws, 44
Armadillos, 23, 27, 28
Artificial selection, 5, 6
Artiodactyls, 32
Atomic theory of matter, 11, 12
*Australopithecus afarensis*, 33, 34, 35

## B

Bacteria, 1, 4
    age of, 21
    antibiotic-resistant, 7, 39
    DNA, 31
    flagellum, 40
    generations per million years, 7
Bats, 24, 27
"Big Bang," 18–19, 20, 51
Birds, 3, 8, 9, 23, 26, 41
Blood circulation and clotting, 12, 41
Bonobos, 35
Brain size, 9, 33–34
Brown University, 15

## C

Cell theory, 11
Central Conference of American Rabbis, 13
Chimpanzees, 24, 30, 31, 33, 35
Chromosomes, 29
Clergy Letter Project, 14
Collins, Francis, 15
Comparative anatomy, 24–26, 30
Computed axial tomography, 38
Copernicus, 18
Coyne, George, 15
Creationism. *See also* Intelligent design creationism
    books on evolution-creationism controversy,
        58–59
    "old Earth," 37–38
    in public schools, xi–xiii, 43–45, 52, 53
    views of proponents, 37–39, 52–53
    "young Earth," 37, 39, 52–53
Cyanobacteria, 21
Cystic fibrosis, 30

## D

Darwin, Charles, xiii, 12, 22–23, 28, 32, 38, 39
Darwin, Erasmus, 22–23
Dinosaurs, 3, 9, 23, 26, 38
Divergence from ancestral form, 8, 9
DNA
    comparisons, 29–31, 33
    defined, 4
    evidence of evolution, xii, xiii, 17, 28–32, 58
    gene duplication, 41
    microarray technology, 5
    mutations, 4, 28
    and recency of common ancestry, 29–31
    sequencing, xii, 28
Dolphins, 25–26, 32
Domestication of species, 5, 6
*Dorudon*, 32
*Drosophila* species, 26, 27, 29, 30

## E

Earth
    age of, 17, 19, 20–21, 38, 51
    origin of, 57
Education. *See* Public school science curriculum
*Edwards v. Aguillard*, 44
*Epperson v. Arkansas*, 44
Evolution. *See also* Humans
    acceptance vs. belief, 49–50
    analogous structures, 25–26
    anatomical homologies, 24–25
    behavioral commonalities, 17, 24, 25
    books on, 56–59
    compatibility with religious faith, xii, 12–15, 49, 54
    continuing nature of, 18, 39
    defined, 4
    DNA evidence, xii, 17, 28–32, 58
    as fact, xiii, 3, 11, 12
    fossil record, xi, xii, xiii, 1–3, 9, 11, 17, 21, 22–23,
        32, 33–35, 38–39, 42–43
    geographic evidence, 17, 26–28
    last common ancestors, 8, 24, 25
    natural selection and, 50
    predictions from findings, 3, 11, 38
    in public school education, xi–xiii, 43–45, 47, 52
    recency of common ancestry, 24, 30
    scientific disputes about mechanisms of, 50–51
    scientific understanding of, xi, xii, 10–12, 39, 49–50
    theory, 3, 11, 12, 51
Evolutionary biology
    agricultural applications, 6
    industrial applications, 9
    medical applications, 5
    practical value, xi, 5–6, 47

*Executive and Managing Editor*
Stephen Mautner

*Sponsoring Editor*
Dick Morris

*Book Design*
Francesca Moghari

*Cover Design*
Michele de la Menardiere

*Photo Research*
Christine Hauser

*Production Manager*
Dorothy Lewis

## PHOTO AND ILLUSTRATION CREDITS

*t* = top; *b* = bottom; *l* = left; *r* = right

**Front Cover** (*tl*) red chrysanthemum, Stockbyte; (*tr*) Monarch butterfly, Don Farrall/Photographer's Choice; (*bl*) Sin Nombre virus particles at 171,000 magnification, Digital Stock; (*br*) orangutan, Ryan McVay; **x** view of Earth, Photodisc; **xvi** flock of shorebirds, Photodisc; **2** (*tl*) Nunavut valley, Ted Daeschler/Academy of Natural Sciences/VIREO; (*bl*) *Tiktaalik* fossil, Shubin Lab, University of Chicago; (*r*) drawing of *Tiktaalik* fin skeleton, Kalliopi Monoyios; **5** airplane passengers, Reuters/Corbis; **6** wheat field, Agricultural Research Service; **7** Trinidadian guppies, Sean Earnshaw and Anne Magurran; **8** loggerhead turtle laying eggs, Lynda Richardson/Corbis; **9** ethanol gas pump, U.S. Department of Energy; **16** meerkats, Photodisc; **19** (*t*) Hubble Space Telescope deep field image of distant galaxies, NASA, ESA, S. Beckwith (STScI), and the HUDF Team; (*br*) Hubble Space Telescope image of planetary disk in foreground of star, D. Padgett (IPAC/Caltech), W. Brandner (IPAC), K. Stapelfeldt (JPL), and NASA; **20** scientists viewing monitor, Agricultural Research Service; **21** stromatolites at Carbla Point, Shark Bay, Western Australia, photo by Jere H. Lipps, February 16, 2002; (*inset*) cyanobacterium *Microcoleus chthonoplastes*, NASA Microbes Image Gallery; **23** transitional bird-like fossil, courtesy Hailu You, Chinese Academy of Geological Sciences; **24** chimpanzee in the wild, Michael D. Carleton, Curator of Mammals, National Museum of Natural History, Smithsonian Institution; **25** (*bl*) dolphins, Captain Budd Christmann, NOAA Corps; (*br*) great white shark, Oxford Scientific/Photolibrary; **26** (*b*) Hawaiian *Drosophila*, Kevin Kaneshiro; **28** nine-banded armadillo, Eric and David Hosking/Corbis; **29** polytene chromosome from a *Drosophila* larva, photograph courtesy Hampton L. Carson, based on an original photograph by Harrison D. Stalker, Washington University, St. Louis; **30** drawing of *Drosophila* and mouse embryos, Sean Carroll; **32** skeleton of *Dorudon*, Philip Gingerich and the University of Michigan; **33** Laetoli footprints, John Reader/Science Photo Library; **36** rainbow heliconia, Photodisc; **40** *H. Pylori* bacterium with flagella, Visuals Unlimited/Corbis; **41** drawing of eyes in living mollusks, reprinted with permission from *Encyclopædia Britannica*, © 2005 by Encyclopædia Britannica, Inc.; **42** Colorado River, National Park Service; **46** diatoms in fresh water, Photodisc; **Back Cover** (*l*) starfish, Slede Prels/Photodisc.